Communication for Engineering Students

D1093466

Communication for Engineering Students

John W. Davies

 LONGMAN

Addison Wesley Longman Limited
Edinburgh Gate, Harlow
Essex CM20 2JE, England
and Associated Companies throughout the world.

© Longman Group Limited 1996

First published 1996
Second impression 1996

British Library Cataloguing in Publication Data
A catalogue entry for this title is available from the British Library.

ISBN 0-582-25648-8

Library of Congress Cataloging-in-Publication Data
A catalog entry for this title is available from the Library of Congress.

Set by 7 in Sabon 10/12 pt
Produced through Longman Malaysia, TCP

Contents

List of checklists

Acknowledgements

I am very grateful to the following people for their advice and help.

At University of Westminster: Paul Regan, Jim Armishaw, Tony Booth, Dennis Preddy, Edin Moossavinejad (Civil Engineering), Ken Bird (Electronic and Manufacturing Systems Engineering), Peter Ormiston (Mechanical Engineering), Colin Everiss (Mechanical Design), David Firth (Manufacturing), Fiona Middleton (librarian), Kathleen Hudson (careers adviser).

At University of Wales Aberystwyth: Frank Bott, Ann Robertson (Software Engineering).

Engineering students and graduates of University of Westminster: Jason Small, Michael Gilby, Thomas Scorer.

Practising engineers: Brian Duncan (civil/offshore), Sheila Fergusson (chemical), Nick Frearson (electronics).

Figure 6.17 has been reproduced with the kind permission of Ms Ying Ying Pang, student on BEng (Hons) Mechanical Design (Engineering), University of Westminster.

1. Introduction

This is a book about written and spoken communication intended specifically for engineering students. Its aim is to help you learn to communicate well, and consequently to become a more successful student and a more effective engineer.

1.1 The importance of communication for engineers

Our word "engineer" can be traced back to the Latin *ingenium* meaning cleverness, or natural ability. The main business of professional engineers is to be ingenious: to come up with good ideas and make them work in practice. No engineer works in complete isolation; there is no point in having a good idea if you are unable to communicate it. Poor communication can create ambiguity, even cause disasters. At the very least it gives a bad impression: if people think you communicate badly, they won't trust you as an engineer. These people could be prospective employers, bosses, colleagues, clients, the public, or the media. There's a great deal at stake. Your career as an engineer, the quality of your achievements, the benefit to society of engineering projects in general, the status and reputation of the whole profession, all these things depend on good communication.

Yet graduate engineers are notoriously poor at communicating. *You* may be an exception of course, but engineering employers widely believe that there is a problem. They also feel that things are getting worse. Typical comments are: "Our graduate engineers cannot write simple letters in decent English", or "I found six spelling mistakes in three lines – I haven't time to correct other people's spelling". As a young engineer you will make progress in your career by taking responsibility; but no matter how good you are at the technical side of the job, if you cannot write a decent letter you won't be left in charge.

Why do graduate engineers tend to be poor at communicating? Let's consider what has happened to you in the last few years. By sixteen you were probably showing promise in maths and science, and since then your education has been quite narrowly based in these areas. English was never your favourite subject and if you were taught grammar in a formal way you thought it was rather boring and have forgotten most of it by now. Am I close? Now you are on an engineering course where most of the challenge and excitement seems to be in the technical subjects.

But let's stop there. You are reading this book, which probably shows that you accept the point that engineers should be good at communicating. Well, now's the time for you to start improving! Most people learn better when they can see the point of what they are learning, so perhaps at this stage in your education you can get more pleasure from learning how to write well than you did when you were at school.

Engineering courses are usually very concentrated – there's plenty to learn. You probably have some classes in communication within your course, but time constraints are such that the topic is a small element compared with the technical subjects. But don't take this as an indication of importance; remember you will not be a successful engineer if you cannot communicate well. The same goes for your career as a student. Let's express it in positive terms: you will be a far more successful student if you can communicate well. You will be required to submit written work from the start of your course, and later some of your most significant assessments will be based on reports.

Improving your ability to communicate well while you are a student is something you must take responsibility for yourself. If you really want to improve, you will. You should take your communication classes seriously, work hard at each written assignment, and use a book like this as a source of help and ideas. Finding out for yourself is an important skill for students and for engineers.

English is a flexible language, constantly capable of accepting new words and expressions. English style and what is considered "correct" are changing steadily. An engineer's letter written in the style of Shakespeare or even Sir Isaac Newton would seem strange and inappropriate to a reader today. Good English is not based on fixed and absolute rules in the way that mathematics is. But like *engineering*, good writing requires judgement, and good judgement requires confidence. You must know the "rules" before you can break them.

Engineers and engineering students should not lack confidence in communication. Good engineering and good communication have much in common: they both require application of knowledge to achieve results, and both call for imagination and pragmatism.

Of course the ability to communicate well can enhance your personal life as much as your professional life. Good communication has a good effect no matter what form it takes.

1.2 How engineers should communicate

There is just one elementary rule of communication for engineers and students: be clear. This book is designed to help you to write and speak clearly.

When engineers write or speak they should have something specific (usually factual and precise) to say. It should be possible to separate *what* is being communicated from *how* it is being communicated. If you are not sure precisely what you want to say you must stop and work it out first. When you are sure, you must know how to communicate it successfully: for the receiver to understand without loss of precision.

A characteristic of communication by engineers is that words are by no means the only medium. Numbers, tables, mathematical expressions, graphs, diagrams and drawings can all enhance communication. Words are only used where they are needed.

Communicating well requires effort and commitment. There are no short cuts, though the advice in this book should help to ensure that your effort is not wasted.

1.3 This book

This book covers the forms of communication that you are likely to use while you are an engineering student. Your course is preparing you for your professional career, and the communication skills you learn as a student will remain of value to you when you are an engineer. Many forms of communication – reports, letters, interviews – are as important to students as they are to professional engineers. However other "professional" communication skills are not needed much by students: writing company memos, chairing meetings, conducting interviews. These are introduced in a short chapter but are not covered in detail. There are plenty of books on professional technical communication, and some are recommended later.

Communication is essential to everyone, and the word has come to mean many things to many people. This book does not cover the technology of communications, engineering drawings (apart from diagrams in reports), management/personnel/corporate communications, advertising, public relations or media studies.

Engineers have many ways of communicating; using language is one of them. I hope the book will be helpful to all English-speaking engineering students, including those who have not spoken English all their lives, but the book cannot be treated as a substitute for a language textbook.

Apart from a dictionary, I hope that this is the only book on writing and speaking that you will need to use regularly while you are a student. It includes basic elements such as spelling, parts of speech, punctuation, grammar and style. These are covered in many other books – I don't claim to do the job better (though I may sometimes do it differently) – my aim is to include all these aspects under one cover, together with their applications by engineering students to lab reports, project reports and so on. In later chapters I suggest other books that you might find useful for reference, and explain why.

I have already suggested that one of the reasons why graduate engineers are poor at communicating is that they did not think seriously enough about the principles of good English when they were at school. For that reason the early part of this book goes right back to basics. Chapter 2 is about words: spelling, meaning and function. Chapter 3 is about sentences and punctuation. I hope you find this coverage useful. Even if you feel it is an insult to your intelligence, you might still consider dipping in to these chapters to boost your confidence – both start with a diagnostic test.

There are several other tests in the book. They are designed to allow you to try out particular skills or put into practice some of my advice. My answers are at the back.

The book is written in an informal style – not the sort of style that you should use when you write a formal report. Also I don't tend to preface my suggestions or advice with comments like "this is only my opinion but . . .". I tend simply to write "you should do this", or "don't do that". I hope you find the suggestions useful.

2. Words

How good is your knowledge of words? Let's start with a test.

Test 2a

1. Is the spelling of each of these words correct or not? If incorrect, give the correct spelling.

   procedure
   proceedings
   superceded
   immeadiately
   maintenance
   necessarily
   sucessfully
   occasionally
   proffessional

2. **phenomenon** – what does this word mean, what is its plural, and from which language is it derived?

3. Give two examples of *adverbs* in context.

4. What is the difference between the words in each of these pairs?

its	it's
principal	principle
affect	effect
practice	practise
imply	infer

5. Explain the meanings of the following words.

   inflammable
   literally
   verbal

continued

Test 2a continued

6.   More spellings. Correct or not? If incorrect, give correct spelling.

language
gauge
gaurdian
begining
parallel
develop
seperate
accomodation
committee

For answers, see page 159.

2.1 Knowledge of words

The aim of this chapter is to improve your basic knowledge of words. English words are fascinating, and the more attention you give them the more interesting they become. The relationship between spelling and pronunciation, for example, can be quite extraordinary. People who have been speaking English all their lives tend to take this for granted (but still make plenty of mistakes). People who are starting to learn English can find it mystifying.

We are engineers and we like numbers. Let's consider English words for numbers.

One, two: that's a bad start, neither is written as it is pronounced.
Three: OK.
Four: not bad, though it doesn't rhyme with *flour* or *tour*.
Five, six: OK.
Seven: not too bad, but it doesn't rhyme with *even*.
Eight: eh? How could anyone guess from that spelling that the word is spoken like *ate*?

Historically, English developed from the language of the Anglo-Saxons. It had many influences, including the incoming languages of the Vikings and the Normans. It was not until printing became common that universal spelling conventions began to emerge. Before that time language had been mostly in spoken form, with different historical influences dominant in different regions. Even while spelling conventions were becoming established, pronunciation was changing under the different influences, so that many of our spellings are effectively out of date: at one time, for example, the k in **know** was sounded. Attempts at creating more logical spelling conventions have generally failed (though a number of words have different American and British spellings).

A good knowledge of words, their spelling, their function, their meaning, and how to use and select them, is essential for an engineer who wants to communicate well. Since words are the basic components of language, improving your knowledge means going right back to basics, perhaps to some extent swallowing your pride. But it's worth it because spelling mistakes can make you look foolish. You should be sure you know how to spell **professional** before you use the word when writing about yourself; unless you are sure how to spell **superseded**, don't write it in red letters across old drawings. Of course you can check words in a dictionary if you have one handy, and that is a very good habit, but you must have a sound knowledge of spelling to know when a simple word needs to be checked. Computer spell-checks can be useful but they have severe limitations as will be considered later. Also there are times when it is simply not practicable for an engineer to use a dictionary or a computer, for example when writing out an urgent warning notice, or sending a fax in a hurry.

2.2 Spelling

As discussed, the spelling of English words can be difficult and unpredictable. There are some "rules" but they all have exceptions. One that you may have heard is "i before e except after c" to which you should add "when the sound is ee", so we have:

	relieve, piece, achieve
after c:	receive, receipt, ceiling
not ee sound:	height, weight, their
exceptions include:	weir, seize

Another useful rule is that if a consonant is doubled, the preceding vowel has a short sound (with the single consonant, a long sound) as in:

writing	written
hoping	hopping
taping	tapping
holy	holly
later	latter

So there is no excuse for: "I think I am quite good at writting English" (written by an engineering student).

The only way to improve your spelling is to take an interest in the way words are spelt. If you *care* about your spelling it will improve.

It must be time for another test.

Test 2b

1. Is the spelling of each of these words correct or not? If incorrect, give the correct spelling.

   independant
   consciencious
   foriegn
   analyse
   diagram
   disasterous

2. Spell the words for the numbers 14 and 40.

3. Is the spelling correct or not? If incorrect, give correct spelling.

   argument
   aweful
   install
   installment
   benefitted
   anomaly

For answers, see page 10.

A list of commonly misspelt words appears on page 10. It includes all the words in Tests 2a and 2b so you can check your answers. Please check thoroughly – I don't want to be accused of perpetuating your spelling mistakes!

If you really want to work hard at your spelling, get a friend to read out these words (or record them yourself on a tape recorder) and see how many you can spell correctly.

Here are some other spellings to watch.

The past tense of lead ("lead from the front") is **led**; the metal with the same pronunciation is **lead**.

When you mislay something you **lose** it; the word **loose** describes a poorly tied knot.

A **programme** (American: program) is a planned sequence; the shorter spelling is used in Britain for a computer **program**.

Similarly a **disc** (American: disk) is a flat circular shape; computers store data on a **disk**.

2.3 Parts of speech

Parts of speech define the function of words. Here are some simple descriptions with examples.

Parts of speech

noun – a word which names a thing or person

   student   laboratory   certificate

pronoun – a word used in the place of a noun

   he   she   it

verb – a word that describes doing or being

   the student *arrived* on time   *is arriving   has arrived*

adjective – a word that gives more information about a noun

   a *clever* student   the *blue* folder

adverb – a word that gives more information about a verb, an adjective or a whole expression

   the student spoke *cleverly*
   this is *extremely* unusual
   obviously this can't be allowed to continue

preposition – a word that describes a connection with a noun

   you walked *into* the party   effects *of* studying

conjunction – a linking word

   Paul *and* Mary
   I've been to the lectures *but* I don't understand them

Commonly misspelt words

Accommodation acknowledge acquire analyse analysis anomaly
apparent appropriate argument awful Bachelor beginning benefited
business Chaos colleague committee conscientious Definitely develop
diagram disastrous Exaggerate Forty fourteen Gauge government
grammar guarantee guard guardian Height Immediately independent
install installation instalment intelligent Language liaise Maintenance
manoeuvre miscellaneous Necessarily Occasionally occurred omitted
Parallel possess procedure proceedings preceding preferred   privilege
professional Queue Simultaneous separate successfully superseded
Unfortunately unnecessary Weir weird

2.4 Words in use

Plurals

Most plurals are formed by adding **s** or **es** to the end of the word, or
changing **y** to **ies**:

word	words
student	students
stress	stresses
library	libraries

Some words do not change:

sheep
cod

Some words derived from Greek or Latin use the plural form of the original
language:

phenomenon	phenomena
criterion	criteria
formula	formulae
radius	radii
maximum	maxima
minimum	minima
stratum	strata

Some words form plurals in other ways:

index	indices (use this in mathematics, indexes for books)
appendix	appendices

matrix	matrices
axis	axes
analysis	analyses

Apostrophe

An apostrophe (') is added to a word to indicate the possessive form or to mark where letters have been omitted.

Possessive

The standard form is **'s** as in

> the engineer's car
> Paul's calculator

With a plural, where an **s** has already been added, the apostrophe comes after the **s**:

> the engineers' team
> all students' best interests

Omitted letters

When we shorten **I did not** to **I didn't** we add an apostrophe to show where a letter has been omitted. In **I can't** the apostrophe marks where two letters have been omitted, and in **I won't** it shows only some of the omissions.

Cases to watch

It's is short for **it is**, whereas the possessive pronoun **its** has no apostrophe: "It's a good design and its advantages are obvious."

Some people try to use apostrophes for unusual plurals like "throughout the 1990's", but there is no need. You can simply write:

> throughout the **1990s**
> all staff have **PhDs**

Capital letter

Nouns which are names or titles usually start with a capital letter:

> Paul   Aberdeen   Faculty of Engineering

Minor words in a long title (**of** above) are not usually given capitals.

People who are unsure about capital letters tend to use them too often. If you want to say that someone has got a job in mechanical engineering you should not use capitals. However if you are saying that you studied Mechanical Engineering at university you should use capitals because it is the title of the course. I did not use a capital for university in the last sentence; that would only be appropriate when referring to a specific university.

Abbreviations and contractions

When Professor is shortened to Prof. it should be given a full stop. That is an abbreviation (letters chopped off the end). When Doctor is shortened to Dr it is not given a full stop. That is because it is a contraction (letters dropped out in the middle). A contraction ends with the last letter of the original word.

Many people do not bother with the full stop after an abbreviation. Others insist that it should be included. You must make your own choice, but remember two things. A full stop after a contraction (Dr, Mr, Ltd) is wrong. And it is often more elegant not to abbreviate at all (Professor, September).

Remember the difference between these two common abbreviations of Latin expressions:

e.g. means **for example**
i.e. means **that is**

Hyphens

Some words include hyphens, for example:

cross-section
self-replicating
de-icer

Many other words which you might expect to need hyphens do not:

reinforcement
semiconductor
preamplifier
coordinates
subroutine

None of this should present any problems. If you want to know if a word has a hyphen, look it up in the dictionary.

Hyphens *between* words are considered as punctuation in 3.2, p 22.

2.5 The right word

There is no point in being able to spell a word or form its plural correctly if it is the wrong word. Engineering students learn a wide range of technical terms yet often seem to have the most problems with everyday words.

You should own a dictionary and get into the habit of using it to check words that you are unsure of. (Many engineering students also seem to own a thesaurus. You may find one useful; see **Further reading** at the end of this chapter.)

Test 2c

What do the following words mean?

nouns:		adjectives:	
	calibration		fallacious
	arbitration		ingenuous
	litigation		tangible
	prose		ephemeral
	enormity		fulsome
	verbosity		obsequious
	pragmatism		implicit
			explicit
			disinterested
			indifferent
verbs:	refute		comprehensive
	precede		comprehensible
			homogeneous
			heterogeneous

Answers: in your dictionary!

Words to watch

Inflammable

This means easily set on fire, from the word inflame. However people tend to associate the in- prefix with a negative (like inappropriate). So to ensure that an important warning is not misunderstood, the words flammable and nonflammable are now common.

Literally

This word is frequently misused. You often hear people say things like "He literally blew his top". This is very hard to imagine! **Literally** means not using a figure of speech. You may be "literally out of your depth" in water, but not when your job is too hard.

Verbal

In the phrase **verbal communication**, verbal means in words, not specifically in speech. Reports and letters, as well as speeches and lectures, are forms of verbal communication. Speech can be called **oral** communication, though the word oral (meaning by mouth) seems rather anatomical; oral communication could include singing, playing the bugle, lip-reading and painting by the mouth. **Spoken** communication is a clearer term.

Difficult pairs of words

fewer/less

This should be an easy one for engineering students – fewer is for integers, less is for other quantities. So:

> **less** than 20%
> **fewer** than 50 qualified engineers

affect/effect

Affect is usually a verb, effect is usually a noun.

> the strike will **affect** deliveries
> the strike is having an **effect**

Both of the following examples from student reports are wrong.

> **How does all this effect students and their chances of getting a job after graduation?**
> **The government's anti-inflationary policies have had a crippling affect on the construction industry.**

complement/compliment

A complement is a thing which fits well or makes something complete. A compliment is a polite expression of praise. Both words can also be verbs. You would complement a team if you brought skills that the team lacked. You compliment people when you say something nice about them. The adjectives complementary/complimentary are also easy to confuse.

discrete/discreet

The first word is associated with mathematics and the second with good manners.

practice/practise

Practice is the noun, practise the verb.

> I want to **practise** engineering, so I have set up a **practice** with my brother.

principal/principle

Principal is usually an adjective meaning first in importance. As a noun it can be used as the title for a head of a college. A principle is a scientific or moral law.

> Among the **principles** of mechanics, engineering students learn about **principal** stress.

plane/plain

Plane refers to a surface, real or imaginary, which is flat; plain means simple, unornamented.

relevant/relative

A common mistake is to write "facts which are **relative** to the discussion" when what is meant is "facts which are **relevant** to the discussion".

continuous/continual

Continuous means without interruption. Continual means regular but not continuous.

> With **continual** distractions you get some work done. With a **continuous** distraction, none.

imply/infer

Experimental results might **imply** something about a material. You, the investigator, would **infer** that it was the case.

alternative/alternate

You may have use of the car on **alternate days** (every other day). When you don't have the car, you have to make **alternative** travel arrangements.

Test 2d

This covers all aspects of the chapter. Spot the eleven mistakes with words (don't worry about the style).

Other information to support my application

My principle reason for applying for this post is that I feel that C G Freeland is a Company which can be proud of it's record in engineering. Your new activities in enviromental control (to which I beleive I can contribute) perfectly compliment your existing specialisms. I am interested in post E15 (Graduate Engineer), but would like to be considered for any alternate vacancies.

My main liesure activity is swiming. I have won county medals on no less than six ocasions, and I represented England at the last commonwealth games.

Answer on p 159.

Further reading

A dictionary

Dictionaries are surprisingly cheap to buy. I suggest you have two. One pocket-sized one to carry around with you, and one medium-sized one to keep at home. The spell-checking facilities of a word processor are not a substitute.

A thesaurus

This is not essential, but many engineering students seem to find one useful (again, cheap to buy). A thesaurus gives, for any particular word, a number of other words with similar or related meanings. There are two formats. One, usually with the title Roget's Thesaurus, lists words in blocks, according to meaning (each word appears once). The other has common words in alphabetical order with a list of synonyms for each.

3. Sentences

Let's start with a test.

Test 3a

Divide this piece into sentences and add other appropriate punctuation.

this is a piece of writing with no punctuation what you must do is insert it you may use commas full stops capital letters paragraphs or any other forms of punctuation that you think might be appropriate the main unit of written english is the sentence sentences can be long or they can be short a sentence really expresses one thought if the thought can be expressed in a brief statement it is quite appropriate for the sentence to be short however some thoughts are more complex and are linked together by words like since and because or but in technical english where clarity is of prime importance there is more danger in long sentences than short ones of course every sentence must contain a verb if it doesnt it is in a manner of speaking too short paragraphs are also very useful for bringing clarity to written english the break between paragraphs provides a definite pause in the text punctuation really matters because it helps to make writing clear people who do not write clearly perhaps because their punctuation is poor make a bad impression they may be excellent at their profession in other respects but if they fail to get good jobs or fail to gain their clients trust they are likely to be disappointed in their careers

Answer on p 159.

3.1 Forming sentences

I think this test shows that punctuation helps to make sense of written words. When we speak, we give our words structure by leaving pauses, changing our tone of voice, or making gestures. Written words rely on punctuation, and the most important element is the forming of the sentences themselves. Unfortunately this is an aspect of writing that many engineering students find difficult.

Each sentence forms a complete statement. When you are reading, you may not learn much until you have read several sentences, but you can pause for a moment after each without feeling that a statement is incomplete.

When James got up this morning

is not a sentence. The statement is incomplete; what happened when he got up this morning?

When James got up this morning he noticed that the postman had already called.

That's better. We might still like to know what was in his mail: perhaps a pools win, or a letter from his former lover? But we can pause; we have read a complete sentence.

Test 3b

Here is a paragraph from a student seminar report. The main problems are in the forming of sentences. What are the mistakes, and how would you correct them (without too much rewriting)? My answer follows, so perhaps you should cover it up.

At universities the importance of teamwork is taught by means of group assignments. Where a group is set a task with a solution that consists of many elements. The elements are divided among the group members so that each member is responsible for one particular part. This type of approach teaches students the idea of responsibility as well as how to be an active member of a team. Since each member must make a useful contribution before the group task can be a success.

Answer to Test 3b

The second and last "sentences" sound wrong because they are not complete statements. The easiest way to correct this would be to join these "sentences" to the ones before, by using commas instead of full stops.

> At universities the importance of teamwork is taught by means of group assignments, where a group is set a task with a solution that consists of many elements. The elements are divided among the group members so that each member is responsible for one particular part. This type of approach teaches students the idea of responsibility as well as how to be an active member of a team, since each member must make a useful contribution before the group task can be a success.

If we did not want to reduce the number of sentences, the original non-sentences could be adjusted to make them into sentences:

> At universities the importance of teamwork is taught by means of group assignments. In these, a group is set a task with a solution that consists of many elements. The elements are divided among the group members so that each member is responsible for one particular part. This type of approach teaches students the idea of responsibility as well as how to be an active member of a team. This is because each member must make a useful contribution before the group task can be a success.

A sentence should contain a verb. That is not a rule, it is a piece of advice. Novelists regularly write sentences without verbs:

I tried the door. Locked. What now?

But in an engineering report a sentence without a verb gives the reader a shock.

However, the presence of a verb does not guarantee that a sentence has been written.

When I got up this morning contains a verb but is not a sentence.

Another example. (Whoops, no verb!) *Here is* another example.

Several deficiencies in the product documentation which have been identified by users.

Well, yes, but what about them? This must either continue so that a complete statement is made, or be rewritten. If it is just required as an introductory statement, the **which** can be removed.

When in doubt, keep your sentences short. Short sentences are generally suitable for communicating engineering information. Don't take risks. Short, correct sentences will make you a clear communicator.

3.2 Punctuation marks

We have the following punctuation marks available to us.

1. Full stop [.]

This marks the end of a sentence.

2. Comma [,]

This is the mildest punctuation mark. It suggests that certain words should be grouped together or a slight pause taken. Here are some possibilities.

(a) a pair of commas

I've used my new calculator, the one with the special functions, to check the calculation.

(b) a pause midway

I've checked my calculations, and Dave thinks you should check yours.

(c) an early pause

However, hard work alone is not enough.

(d) a list

We have completed the calculations, the drawings, the model and the report.

Commas would always be used in (d); in (a), (b) and (c) they are helpful rather than necessary. In (a), **the one with the special functions** is a separate phrase describing the calculator, and is marked off by a pair of commas. Without the commas it would be odd to see the two nouns **my new calculator the one** together, and someone might think that the calculator had special functions capable of checking any calculation: **the one with the special functions to check the calculation.** In (b) the comma saves us from momentarily thinking that two things might have been checked: **my calculations and . . .** In (c) the comma creates a dramatic pause and prevents any premature assumption that the sentence starts **However hard . . .**

In novels and newspapers, writers may economise on commas in order to step up the pace. In engineering, clarity is the aim. If a comma helps to make your writing clear, put it in, but make sure it is in the right place. (There are more examples later.)

3. Semicolon [;]

This is a mild full stop. You might write:

I have checked the calculations. Now we can work on the model.

These are two satisfactory sentences. But they are strongly linked to each other. You could write instead:

I have checked the calculations; now we can work on the model.

So a semicolon links two statements, that *could* be separate sentences, into one sentence.

4. Colon [:]

A colon makes me think of a compère or Master of Ceremonies. "Introducing our special guest for tonight: Billy Connolly." The colon marks the moment when the compère makes extravagant arm gestures and a spotlight flashes across the stage. A colon provides an introduction.
Here is an obvious example.

You should bring the following items: boots, anorak . . .

Or

Some things are essential: passport, travellers cheques . . .

A colon could introduce a result or consequence.

I have checked the calculations: they are correct.

5. Paragraph

A paragraph is a group of statements. A new paragraph contains a new topic or a new line of thinking. The break between them is a definite pause. It is not wrong to write a paragraph which contains only one sentence, but if you do it frequently you may be wasting the potential of punctuation to give structure to your writing.

6. Brackets [()]

These isolate a phrase in the same way as a pair of commas (2(a)), but create a stronger separation. They are often used to suggest a change of tone: the remarks in brackets may be more personal or lighthearted (or less important). A sentence should still make sense if the expression in brackets is removed. (A whole sentence or group of sentences can also be written in brackets. In this case, the last full stop comes *inside* the closing bracket.)

7. Dash [–]

Dashes are informal – you probably put lots of them in letters to friends. They usually take the place of colons or commas. They can be used in reports to create informality, but there is no situation in which they are definitely required.

8. Hyphen [-]

We have already considered hyphens within words (in 2.4, p 12). A hyphen can also be used optionally as a punctuation mark between words. Many people seem to be unaware of the usefulness of this option.
The sentence:

We have planned two day long meetings.

would be made clearer by using a hyphen:

We have planned two day-long meetings.

Here are some other examples:

long-running dispute
radial-flow turbine

In another context we might write:

vanes create radial flow at entry

The hyphen is not there now because it is not needed to add clarity.

9. Inverted commas [" "]

These have a number of uses.

(a) Quoting directly from a source

As Shakespeare wrote: "Life's but a walking shadow".

(b) Drawing attention to an expression which should not be taken literally. I referred to **"sentences"** after Test 3b because they looked like sentences but they had not been properly formed. In dynamic testing of a model in a laboratory you might refer to **"earthquake conditions"**. But you should be careful. Engineers must not take risks with the meanings of words. Never write things like:

This "proves" that the theory is correct.

(c) Recording dialogue – not normally needed by engineers.

10. Question mark [?]

This must be used at the end of a question.

What do these results show?

11. Exclamation mark [!]

This is not often needed when engineers write, though it may frequently be implied when they speak!

Test 3c

Simple misuses of punctuation. Identify the mistakes.

1. The number of members within a team depends on two factors; the size and complexity of the project.

2. A quality management system should be based on existing systems; amended and supplemented where necessary to conform with BS 5750.

3. Control should be exercised throughout the whole process from start to finish, products within a subcontractor's work may have to be included.

4. What are the main problems with the current system.

Answers on p 160.

3.3 Sentences and punctuation

Let us now consider how these things fit together. We will look at some more examples, and discuss some further details.

Here are some extracts from student reports, with comments on punctuation. I have not called this a test, but why not cover up my comments first and make a decision yourself?

(1) **An engineer is appointed by a client to assess whether a project is appropriate, at a later date if the project goes ahead, the engineer may be involved a great deal with the technical aspects of production.**

Comments: **at a later date** is the start of a separate statement, and should therefore be the start of a new sentence. The last comma feels as if it should be the second of a pair, isolating **if the project goes ahead**. So I would change this to:

An engineer is appointed by a client to assess whether a project is appropriate. At a later date, if the project goes ahead, the engineer may be involved a great deal with the technical aspects of production.

(2) **It was decided after analysing the subject title, that we should concentrate our efforts on WHY it was important to know who's who in the management of a project, rather than describing, in greater detail the roles of the various members of the team.**

Comments: The problems are with incomplete pairs of commas. The expressions that should be isolated are **after analysing the subject title** and **in greater detail**. That would give:

It was decided, after analysing the subject title, that we should concentrate our efforts on WHY it was important to know who's who in the management of a project, rather than describing, in greater detail, the roles of the various members of the team.

The sentence now has a lot of commas, and it would probably be worth rearranging the last part to give:

It was decided, after analysing the subject title, that we should concentrate our efforts on WHY it was important to know who's who in the management of a project, rather than describing the roles of the various members of the team in greater detail.

(3) **Four months later, the contractors, reported that they were down to their last £1m. In other words, they were broke. Which stopped the banks from paying any additional cash, until they had complete assurance on the costs of completing the tunnel.**

Comments: The comma after **the contractors** is wrong. Also the last "sentence" is not appropriate. It would not make a complete statement if you read it by itself. It could be attached to the sentence before by changing the full stop after **broke** to a comma. But then the second sentence would be a complicated one, and we would have spoiled my favourite bit, the sentence **In other words, they were broke.** So I would keep the last sentence separate with a slight rewording. Finally I don't think the last comma is helpful.

> Four months later, the contractors reported that they were down to their last £1m. In other words, they were broke. This stopped the banks from paying any additional cash until they had complete assurance on the costs of completing the tunnel.

(4) **We must educate people, and change their attitude to engineering. If we fail to do so, there will be no one to blame, but ourselves.**

Comments: The last comma should be removed. It is possible that a politician giving a speech might leave a long pause before saying **but ourselves.** But that sort of effect cannot be achieved in writing.

Test 3d

Here are some short ones for you to sort out.

1. The electronics industry has been healthy compared with other industries this can be clearly seen in the attached graphs.

2. This, coupled with high interest rates has caused many small engineering firms to fold.

3. Engineering will continue to be misunderstood, and we graduate engineers, are the ones who will suffer most.

4. The wall is relatively thin but, it is strengthened at regular intervals by buttress supports.

5. Although these machines rarely need maintenance, do not have regular breaks like their human counterparts and do not arrive late they were not developed to replace humans.

6. Another interesting idea, is one that is currently used in Houston Texas.

7. There is no requirement for the engineer to be present, isn't this unsatisfactory.

Answer on p 160.

3.4 Further punctuation details

The last comma in a list

In a simple list like **Paul, Mary, Mark and Jo** it would be unnecessary and odd to place a comma after **Mark**. Now here is a more complicated list.

> **For each overflow, I carried out a thorough survey on site, made detailed design calculations, prepared a plan and section, and supervised the completion of the contract drawings.**

The comma after **section** helps to make this list clear. In a complicated list, a comma after the second-to-last item is usually a good idea.

Long quotations

Here is an imaginary quotation which explains the point.

> "These sentences have been copied word for word from a source. There could be a number of reasons for quoting them: they might form a famous passage from a book, or the quoted author might have made a point particularly clearly. The quotation is correctly given in inverted commas.
>
> "The quotation is more than one paragraph in length. Each new paragraph begins with inverted commas (to remind the reader that the quotation is continuing). It is only the last paragraph that ends with inverted commas. And these last inverted commas are placed after the full stop because the whole of the last sentence is part of the quotation."

Further reading

See Chapter 4.

4. Grammar and style

4.1 The need for judgement

When people say "That's bad English" they usually mean one of two things. They either mean that the grammar is incorrect: that one of the rules of the English language has been broken; or they mean that the style is poor: that, for example, the writing is so wordy, or full of jargon, that it is hard to understand.

Things can be more complicated. They could be referring to something like a split infinitive, which they might insist was a point of grammar but someone else would dismiss as a matter of taste. I call this third category word-problems.

So good writing requires an understanding of rules, of matters of taste, and of the difficult area where the two overlap. Good taste is subjective; there are commonly held opinions but these change with time. There is no one way of writing well. Good writing, above all, requires judgement, and judgement must be based on knowledge and awareness.

This chapter contains important points about grammar, style and word-problems. It is intended to help you write good English. It may also help you defend yourself against people who criticise your English without justification.

4.2 Grammar

It is sometimes hard for people who have been speaking English all their lives to explain to those who are learning the language why a particular phrase is grammatically incorrect, and another correct. Native language speakers know how to use hundreds of principles of grammar without necessarily knowing how to define or explain them.

If this unconscious knowledge meant that English speakers never made grammatical mistakes, there would be no need for this section. In fact it is easy to make mistakes, and it is often the simplest grammatical principles that cause problems. We will consider some of these now.

Subject, object, pronouns

Subject–verb–object is a simple structure for a sentence. The subject and object are nouns or pronouns or phrases which include them.

In **Paul hit Jim** Paul is the subject, hit is the verb, and Jim is the object. We infer this from the order of the words. **Jim hit Paul** contains the same three words but the effect is very different (on the two people).

The only words that actually change their form according to whether they are subject or object are pronouns:

> *I* hit Jim
> Jim hit *me*
> *They* saw *us*
> *We* saw *them*

Pronouns also change when they follow a preposition.

> *I* gave the book to *her*
> *She* gave the book to *me*

The verb **to be** does not take an object. Consider the sentence **I am an engineer**. Here **an engineer** is called the complement. When a pronoun is the complement it has the same form as it does when it is the subject.

So when you are asked, after you have knocked on the bathroom door, "who is it?", it is grammatically correct to answer "it is I". Of course nearly everyone actually says "it's me". When writing however you should try to be grammatically correct, for example:

The engineers responsible for answering queries are David and I.

People are often worried about sounding pompous when referring to themselves, and **David and me** somehow sounds more modest – but it is wrong.

In informal speech it can feel more comfortable to say things like **Dave and me will answer queries** or me and Dave will cope with the hassle. Here **me** is part of the subject and so is obviously ungrammatical. We would never say **me am going out for lunch**.

People who are trying hard not to say **and me** fall into the trap of saying **between you and I**. This is wrong because **between** is a preposition; it should be **between you and me**. (We would not say **I can't let you come between I and my friends**.) Similarly **leave all the queries to David and me** is correct; **to David and I** would be incorrect.

The boss has nominated David and me to deal with queries is correct because **David and me** is the object. It seems to be the **David and** that puts us off. When in doubt remove other names and work out if **I** or **me** would be correct by itself.

Here is a summary.

Correct
David and I will deal with queries (subject)
The boss has nominated David and me to deal with queries (object)
Leave all queries to David and me (after preposition)
The engineers responsible for answering queries are David and I
(complement)

Singular and plural with verbs

It is easy to make a slip like

one of the many rules of the English language have been broken.

It should be **has** because the subject is **one.**
It is also easy to write something like

**in-process inspection and testing now involves more attention to the
documented quality plan.**

Here the **and** makes the subject plural, so the verb should be **involve.**
However

**the conduct of in-process inspection and testing now involves more
attention to the documented quality plan**

would be correct because the subject is singular, **the conduct.**
The following sentence is correct.

**Lady Windermere, accompanied by her husband, is going to visit the
plant next week.**

The subject is singular: **Lady Windermere.**
It would of course be

Lord and Lady Windermere *are* going to visit the plant.

4.3 Style

The main principle is **be clear.** Engineers achieve clarity in a number of
ways. We will look at the use of numbers and diagrams in Chapter 6. Here
we concentrate on clear use of words. There are no rules, only guiding
principles. There is no one way of writing clearly. It is not an easy thing to
do, but it is worth the effort. You must practise, and always think hard
about what you write.

Here are some ideas you should bear in mind. I will list them first, then give some explanation and examples.

Style points

always be precise
keep it brief
keep it simple
be yourself
make sure it sounds right
be careful with: made-up words
metaphors
clichés
jargon

Always be precise

Engineers must be precise when they use words just as much as when they use numbers. Technical words must be used according to their established definition. Everyday words must be selected to convey a clear meaning.

Keep it brief

If you don't like writing you may welcome the idea that engineers should use no more words than are needed to make their meaning clear. However, writing concisely usually takes more work than writing at length. Keeping it brief requires care and concentration. You can start by avoiding unnecessary words.

The words in italic below could simply be removed.

fewer *in number*
an *approximate* estimate
each *and every* person

It's a good idea to ration your use of **very**. It is not a precise word. You should never write that a material is very strong or a process very energy efficient. You should quantify, or at least compare with a useful reference. **Very** does not normally add anything. You may write in a letter that the delay in receiving the drawings is causing a very expensive delay, but your letter is not made more convincing by the **very** (especially if it includes more than one).

Avoid language which is rambling and vague (possibly meant to sound important).

While the ethos of quality may already permeate an organisation, management should approach the issues of quality assurance with not a little caution, while recognising the need ultimately to address them.

Engineering students do not normally write like this, but some engineers do. If you detect any such tendencies entering your writing, try to prevent them from taking hold.

Keep it brief. Enough said.

Keep it simple

Avoid long words. Keep your sentences short. Use straightforward language. Sometimes there can be a conflict between keeping it brief and keeping it simple. In these cases, simplicity should have priority.

The software contains global input default parameter values to facilitate initial model building.

This is brief, but it is not simple. The technique of loading adjectives on to nouns (**values** has four) produces concentrated but difficult writing. Using nouns as adjectives (input, parameter and model above) has the same effect. People cannot accept information if it is delivered too quickly, so don't try to pack too much into one sentence.

Be yourself

Have the confidence to write the way you think is best. Don't copy other people's bad English. When you start work in industry, don't try to adopt a special "business style" – there is no such thing. The aim is constant: be clear.

Make sure it sounds right

Sometimes, when you check over a sentence that you have written, it seems wrong. You cannot work out why. It is grammatical, precise and simple, but it doesn't sound right. There could be all sorts of reasons: word order, for example, or too much repetition of a particular word in the piece as a whole. You are not expected to be a poet, but if something doesn't sound right to you it probably won't to your readers. You should try to find a way of improving it.

Be careful with made-up words

English has plenty of scope for making up words – deriving adjectives from nouns, verbs from adjectives and so on.

If, on your course, you have to study a fixed number of standard-sized **modules** (noun), then your course is **modular** (adjective). To make a course modular, someone must **modularise** it (verb). When that process is finished the course might be described as recently **modularised** (adjective). The process is **modularisation** (noun). Five years later the institution will question the benefits of **modularity** (noun: the condition, rather than the process of change).

When using these sorts of words, make sure that you use them correctly, and bear in mind that there is no need to use an ugly word when a simple expression will do. I suppose the word "modularisable", if it existed, would mean capable of being modularised – fortunately it doesn't exist.

Metaphors

A metaphor is a word used for effect in a way in which its meaning cannot be taken literally:

there are new projects in the pipeline
he is a giant in his field
it was a stormy meeting

Many are so common that we tend to forget that they are metaphors: **impact, target, ceiling, kept in the dark.** This creates the danger that we use them in an inappropriate way:

representatives are feeling snowed-under by the lack of coherent strategy.

And we can end up with some quite laughable statements:

new water supply projects are in the pipeline
supplies of blood are drying up this summer
ferry companies are taking on board new safety proposals

(also see Clichés, next).

Clichés

A cliché is an expression which might have been strikingly effective, even amusing, when first used, but has since become stale and worn-out. The person who first said "all we ask for is a level playing-field" probably inspired a review of procedures for competition between firms. Now the expression is badly worn and slightly irritating. Unfortunately many engineers, in speech at least, seem to be fond of clichés; it is almost as if they feel that these are the expressions that *should* be used.

If you care about the way you communicate you should try to avoid clichés, especially in writing. If you want to amuse, think up a fresh expression of your own. Otherwise simply say **fair competition** rather than **level playing field**, and **rough estimate** rather than **ballpark figure**. Build up your own list of clichés to avoid. Here are some more to get you started:

take on board
moving the goal posts
at the end of the day

Jargon

When engineers have the opportunity to explain their work to non-engineers they should express themselves in a way that can be understood. That's obvious. Engineers have a reputation for communicating badly to a wider audience, and that is a great shame. If a student friend, a non-engineer, asks you in the pub about your final year project, don't say "oh you wouldn't understand"; try to explain, in everyday language, what you are doing. The low level of public knowledge of engineering is partly the fault of engineers.

If engineers try to communicate with the public in language that would only be understood by engineers, they are using jargon. However, when they communicate with other engineers they should make full use of the common technical language.

4.4 Word-problems

I have given some rules for grammar, and some guidelines for style. Now we come to points of English usage about which people disagree. I will give you background information, and in some cases my opinion. You must make up your own mind.

Split infinitive

An infinitive is a verb in the form: to go, to walk. It is split if a word is placed between the **to** and the verb: to boldly go, to quickly walk. Split infinitives often sound ugly, and in those cases you should avoid them. But most experts on English usage agree that it is a matter of taste not of grammar.

He or she

It used to be normal to write

> anyone who cares about the way **he** speaks
> someone who has been speaking English all **his** life.

It was a convention; it didn't mean that the comment only applied to men. If someone today writes

> **An engineer must have confidence in his decisions**

readers will infer, rightly or wrongly, that the writer has forgotten that the engineer might easily be a woman. It is so important not to give that impression that this sort of wording should not be used. The problem can be solved by writing

> **An engineer must have confidence in his or her decisions**

But suppose the piece continued like this:

> **An engineer must have confidence in his or her decisions. He or she may have to defend his or her decisions to the public, to his or her employer or his or her client. These are the responsibilities that he or she has entrusted in him or her by virtue of his or her position in society.**

Things can get clumsy! It could be reworded of course, but the simplest solution might be to put it in the plural.

> **Engineers must have confidence in their decisions**

If we are going to avoid **he** with **engineer**, we might as well be consistent. It would be possible to change our first example to

> **anyone who cares about the way he or she speaks**

but again we risk clumsiness. In the search for a solution, some people come up with

> **anyone who cares about the way they speak**

but that is ungrammatical. Why not make it plural and do the job properly?

> **people who care about the way they speak**

Data

The plural of the Latin word **datum** is **data**. So some people think **data** should be treated as a plural in English:

> **the data have been analysed**

But English has other plural Latin words that are treated as singular: agenda, stamina. Moreover in current English usage data is not the plural of datum. A datum is a fixed reference point, data is information (often in number form).

The modern way, for students and engineers, and people who work with computers, is to use data as singular:

the data has been analysed.

Informality

Contractions like **don't** and **can't** are informal. Most reports by engineers and engineering students tend to be formal, and unless you want to create a particular effect you should write **do not, cannot** etc. (This book is written informally.)

Just as formal expressions are out of place when chatting with friends, colloquial expressions are out of place in most reports. Phrases like those below do not lighten the tone, they simply stick out like a sore thumb.

stick out like a sore thumb
there is no way that . . .
it is a rip-off
the experiment was a total disaster

Further reading

Gowers, Sir Ernest (revised Sidney Greenbaum and Janet Whitcut) *The Complete Plain Words*. Penguin, 1987. This is probably the most popular British book on clear simple English. It was originally written to encourage civil servants not to be pompous and wordy, to replace "This office is in receipt of your communication" by "Thank you for your letter". It is full of excellent advice.

Dummett, Michael *Grammar and Style*. Duckworth, 1993. The author is very intelligent and very fussy. If someone has accused you of making a grammatical error, this book would be a good one to use to settle the argument. However, thoroughness not simplicity is the author's main aim, and the book is not suitable for anyone lacking confidence in writing English.

Bryson, Bill *Dictionary of Troublesome Words*. Penguin, 1987. This is a lively reference book which concentrates on word-problems, with troublesome words set out alphabetically.

You may find that it is worth buying a pocket-sized book on English usage. There are several on the market, no more expensive than a small dictionary or thesaurus.

5. The writing process

Everything you write will involve some use of the procedures that are considered in this chapter. Later parts of the book refer back to this chapter more than to any other.

5.1 The blank sheet of paper

Suppose you need to write something – an important letter, a coursework report. You must stop to think before you write, so you get some paper and think. How will you start? Think. Perhaps you will begin by explaining the problem? Hmm, think. Make a cup of coffee. Think. Close the window. Think. Think. STOP!

The problem of getting started, the fear of the blank page, is a common experience. Some people call it "writer's block", but that is a misleading expression. Fiction writers may experience writer's block – a lack of creative ideas, nothing to say. But engineers and engineering students usually have something to say, otherwise there would be no point in setting out to write. The problem is putting it into words. But don't worry – the techniques are easy to master. There is *no need* to stare at a blank sheet.

There are three stages: defining the task, sorting out ideas, and putting ideas into words. Let's consider them separately. But first remember one important thing. Everybody is different. Everyone gets ideas in a different way, thinks in a different way, and writes in a different way. Different techniques work for different people. The same person may even use different approaches for different types of writing. You must work out what is best for you. The ideas that follow should help.

5.2 Defining the task

You must: define your subject precisely
define your aim in writing about it
define your readership.

This should not be difficult for a short student assignment. Your subject has probably been defined for you, though this is not always the case. Your aim is to learn, and to score marks. Your readership is your lecturer or lecturers, and possibly your fellow students. For a longer student report or a professional engineer's report this defining stage is more significant, and it is considered in more detail in Chapter 10 (Reports).

5.3 Sorting out ideas

You have defined the task. You might have made some notes of your definition, or simply written the title at the top of the page. Now you must collect and sort out the ideas upon which your writing will be based. I think there are three stages to this:

collecting ideas
sorting them out
creating a structure.

Ideas

The types of ideas that will go into your writing depend entirely on what you are writing about. You may have been researching a topic in the library, taking measurements in the laboratory, carrying out design calculations, or testing a piece of software. You may have plenty of information: your desk may be piled high with books, notes and graphs. But how do you write about it?

The answer is that at first you put writing out of your mind. You take a sheet of paper and scribble down ideas. Let nothing disturb or inhibit the flow of ideas – don't worry about how you will write about them, how you will relate one point to another, or even whether all the ideas are relevant. Just write them down. Keep this going as long as you can. You may, depending on time constraints, be able to do this over more than one session. For a major report you might even be able to jot down ideas, when they come to you, over a period of weeks.

Sorting

When the ideas have stopped flowing, you must sort them out. This is likely to mean grouping ideas together, finding themes, and probably crossing some ideas out. As you sort out your ideas, you may be able to add some new ones.

Structuring

Only now should you think about the piece you are going to write. The groups of ideas must be structured so as to communicate them most clearly to the reader. You must consider sequence, titles for different sections, and perhaps a numbering system (more about that in Chapter 10).

---

employers' comments about communication skills – why this situation

how to use book – could skip 2 and 3 or use as confidence booster, each starts with test

no point in having a good idea if you can't communicate it

no engineer works in complete isolation

in degree, not much time for communication

bad English could: cause ambiguity, disasters, or give bad impression

English style is changing – engineer's letter in style of Shakespeare would sound strange

meanings of "communication" – what's covered/not

if people don't like your English, they won't trust your engineering – prospective employers, bosses, clients, the public, the media

some words and forms are changing – not like maths – but like engineering, good writing requires judgement

like engineering – you've got to know "rules" before you can break them

book covers forms of communication needed by students while they are students – of course degrees are a preparation for a professional career and the communication skills learnt will also be valuable then

engineering students/graduates are not good (generally) at writing (speaking)

be clear

origin of the word "engineer"

use of English, applications, it's all in here – recommended books

importance of finding out for yourself

school – whether or not taught grammar, you've probably forgotten most of it, or didn't think it was important

English – absorbs new words

now's the time to learn to communicate well

back to basics (but can skip)

Figure 5.1 (a) Ideas

Figure 5.1 is based on my own notes for writing Chapter 1 of this book, in the separate stages of **ideas, sorting** and **structuring.**

communication is important for engineers – why

graduate engineers are known to be poor communicators

maybe you didn't learn much grammar at school – now's a good time to learn

communication does not have rules like maths – but that's like engineering

scope of book

it's all in here (recommended books)

English changing, absorbing words

may not spend long on Communication in engineering course – must learn for yourself

be clear

word "engineer"

how to use book

Figure 5.1 (b) Sorting

1.1   *The importance of communication for engineers*
          word "engineer"
          why communication important for engineers
          graduate engineers known to be poor communicators
          maybe didn't learn much at school – now's a good time
          may not spend long in engineering course – must learn for yourself
          English changing – communication does not have rules like maths – that's
             like engineering

1.2   *How engineers should communicate*
          be clear

1.3   *This book*
          scope
          it's all in here (recommended books)
          how to use

Figure 5.1 (c) Structuring

5.4 Putting ideas into words

Putting it all into words also has three stages:

preparing
writing
improving.

Preparing

Your ideas have been sorted and structured. Now you start work on putting them into words – one section at a time. Each idea may still be represented by only a few words. You may need to expand this in note form so that you are clear about all the points you wish to make. My own **preparing** notes for part of Chapter 1 are given on Figure 5.2.

Writing

You have prepared so carefully that now the piece should write itself. No? Then perhaps you have not prepared carefully enough. Writing is not a magic process in which words are created from nothing. You should not try to write if you do not know exactly what you want to say. If you get stuck, don't try to go on, go back to your notes. Step back a stage, from writing to preparing. There is no need to step forward again until you are ready.

When the words finally start to come, don't be too fussy. Remember that you are only producing the first draft. Don't try to get everything perfect. Even if you are quite unhappy with some sections, it is usually best to carry on and leave improvements until later.

Improving

Writing of the first draft is now complete, and you can congratulate yourself. You may suspect that what you have written is far from perfect, but you have overcome the big obstacle – you have written. The job is by no means finished; the next stage, **improving**, is one of the most important. You should consider everything carefully, including English, factual content and structure. You should read at normal speed to see if the writing is clear. You should read slowly to check the details. Ideally you should carry out a complete set of revisions, leave the piece for a few days, then go back and check it again. Another way of getting a fresh view is to persuade a friend to check through it.

1.1

(a) communication is important for engineers – why

no point in having a good idea if you can't communicate it – no engineer works in complete isolation (link to word "engineer")

bad English could: cause disasters/ambiguity or give bad impression

if people don't like your English, they won't trust your engineering – prospective employers, bosses, clients, the public, the media

(b) graduate engineers known to be poor communicators

employers comments about comm skills

(c) why poor?

you preferred Maths to English, narrow education, maybe didn't learn much grammar

school – whether or not taught grammar, you've probably forgotten it now, or didn't think it was important

Figure 5.2 Preparing

The stages of the writing process are summarised on Figure 5.3.

Everything you write is different

You will not need to follow all these steps, in the same way, for everything you write. On a good day, your ideas may come to you already sorted. Or you may be so clear about what you want to say that you can start writing as soon as you have finished structuring your ideas.

The **preparing** stage is the one that is likely to vary the most. If you are finding it easy to write something, you will need to spend less time preparing. If you are finding it difficult, you will need to spend more.

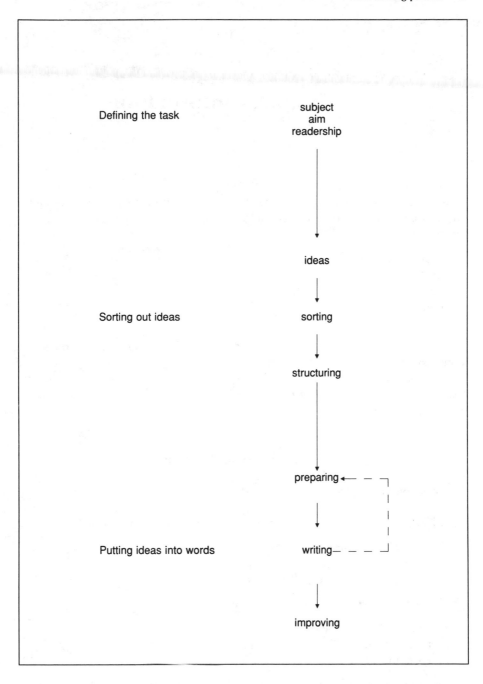

Figure 5.3 The writing process

6. Technical information

Engineers and engineering students do not communicate with words alone. In an engineering report, technical information is presented using a *partnership* of:

numbers and symbols
tables
graphs
diagrams
. . . and words.

This may sound like the familiar content of an engineering course (in contrast with the first few chapters), and certainly not all of the advice in this chapter will be new to you. The aim is to bring together guidelines for using these partners, not just to record and analyse technical information, but to *communicate* it.

6.1 Numbers and symbols

How to write numbers

Small integer numbers without units are normally written as words:

three alternative designs were studied
there are **six** main reasons for proposing this scheme
students study **eight** modules in each year.

Otherwise numbers are written as figures:

there are **146** possible error messages
a current of **3.7** mA
at **25** m intervals

Large numbers may be made easier to read by leaving a space between each group of three digits:

46 230   23 200 000

The traditional British practice of using a comma in this space is potentially confusing since a comma is used as a decimal point in many countries.

It is often helpful to write large or small numbers in the form:

$$2.69 \times 10^6 \qquad 7.03 \times 10^{-3}.$$

However the choice of units (considered later) may make this unnecessary.

Accuracy

When results of measurements or calculations are given, you must be very careful about accuracy.

Strictly, 0.5 means a value between 0.45 and 0.54. 0.500 means between 0.4995 and 0.5004.

You may, after careful consideration of accuracy, have written down a measured quantity as 2.63 (units will be discussed later). In calculation you may divide this by another measured quantity, say 4.81. Your calculator will probably give the answer as 0.5467775; but if you include all those decimal places when you quote the result in a report, you are not only being unscientific, you are communicating badly. The result of the calculation cannot be more accurate than the measurements upon which it was based. The way you write a number communicates its accuracy.

Units

You should use standard SI units and their standard abbreviations. The base SI units are:

metre	m	kelvin	K
kilogram	kg	candela	cd
second	s	mole	mol
ampere	A		

Abbreviations for other common engineering SI units are:

radian	rad	volt	V
hertz	Hz	farad	F
newton	N	ohm	Ω
pascal	Pa	siemens	S
joule	J	weber	Wb
watt	W	tesla	T
coulomb	C	henry	H

Other common engineering units, not strictly SI, are:

angle:	degree	°
	minute	′
	second	″

hectare	ha
tonne	t
litre	l

time:	day	d
	hour	h
	minute	min

Since these are standard abbreviations, it is obviously important that you use them and do not make up abbreviations of your own. There are no plural forms; **m** is the abbreviation for metre and metres. (My Physics teacher at school used to say that we were "far too fond of **secs**".)

Prefixes are used for multiples of these units, for example a kilometre, km, is 10^3 metres. The common prefixes are:

Abbreviation	Multiple
kilo	k	10^3
mega	M	10^6
giga	G	10^9
milli	m	10^{-3}
micro	μ	10^{-6}
nano	n	10^{-9}

Of course many units are made up of combinations of these units. For example velocity is distance divided by time, metres per second. Now we hit a problem. A scientist would write this as **m s⁻¹**, but most practising engineers would write **m/s**. Some engineering textbooks have units written one way, other books on the same subject have them written the other way. Lecturers' views also differ. While you are a student it is probably best to fit in with other people. There is no point in writing units one way when your lecturer and textbook both use the other way.

You should leave a space between the quantity and the unit:

6 mm	6.92 mm	930 kg/m³
11 kV	0.57 m/s	25 N/mm²

Symbols

Each branch of engineering has a fairly standard set of symbols. You will communicate most clearly if you follow the conventions. Everything you write must be consistent in use of symbols, and all the symbols that you use must be defined. If a report contains more than about ten symbols, it is worth giving a separate list of symbols in which they are all defined together.

Many symbols used in engineering science are Greek letters. How do you say the word for ζ or ξ? In case you've always been afraid to ask, here is a full set of Greek letters (capitals second). Now you can impress your friends with your knowledge of the Greek alphabet, *and* work out the destinations of buses when you are on holiday in Greece (though I don't suggest you take this book to read on the beach).

Greek alphabet

α	A	alpha	ν	N	nu
β	B	beta	ξ	Ξ	xi
γ	Γ	gamma	o	O	omicron
δ	Δ	delta	π	Π	pi
ε	E	epsilon	ρ	P	rho
ζ	Z	zeta	σ	Σ	sigma
η	H	eta	τ	T	tau
θ	Θ	theta	υ	Y	upsilon
ι	I	iota	φ	Φ	phi
κ	K	kappa	χ	X	chi
λ	Λ	lambda	ψ	Ψ	psi
μ	M	mu	ω	Ω	omega

Equations

All equations should be written on separate lines, as clearly as possible. Each equation should be numbered in brackets on the right side of the page, and then referred to by that number.

$$H = \frac{p}{\rho g} + \frac{u^2}{2g} + z \tag{6}$$

Equation (6) is used to determine the total head . . .

If the symbols have not been defined in a separate list, each new symbol must be defined as soon as it is introduced. In this case, immediately under equation (6) we might write:

where H = total head (m)
$\quad\quad p$ = pressure (N/m^2)
$\quad\quad u$ = velocity (m/s)
$\quad\quad z$ = vertical distance above datum (m)

In a long report which is divided into numbered sections, equations in Section 2 should be numbered (2.1), (2.2) etc.

Statistics

The sensible use of statistics can be of great value in making precise comments about data. Wherever you can, you should give a statistical parameter instead of using vague phrases like "reasonable agreement" or "similar to previous values". An understanding of statistics is essential when presenting technical data of any complexity.

Presenting statistical information clearly and objectively can sometimes be as much a test of communication skills as of mathematical understanding, as demonstrated by Darrell Huff's book, "How to lie with statistics" (see **Further reading**).

6.2 Tables

A table is a neat way of presenting a collection of information. In an engineering report this information is often in the form of numbers. Values of each parameter are set out in the columns of the table. Each column has a heading giving a brief definition of what it contains, with a symbol and units if appropriate. Ideally tables should be kept as simple as possible, and they shouldn't contain unnecessary columns. There is no need for a column that shows an intermediate stage in a simple calculation, and certainly no need for a column in which all the numbers are the same. The column on the left may identify each row by stating, for example, the value of the independent variable, or the reference number of the specimen.

All comments in the previous section about accuracy and units apply to tables. The numbers in each column should be arranged so that decimal points are in a vertical line.

In some cases you may have a column of very large or very small numbers like this:

Volume stored
S
(m^3)

5.3×10^6
9.1×10^6
16.8×10^6
21.0×10^6

Let's assume that it would be inappropriate to change the units. Writing $\times 10^6$ so many times is a nuisance and it would be convenient to write it just once, at the top of the column. You must be careful. Some people write the column heading like this:

Volume stored
$S \times 10^6$
(m^3)

5.3
9.1
16.8
21.0

But this is wrong. It implies that the column contains values of S which have already been multiplied by 10^6 (not which still need to be multiplied by 10^6).

To convey what you really mean you should write the column heading like this:

Volume stored
S
$(m^3 \times 10^6)$

5.3
9.1 etc

This says, in effect, that each number represents so many millions of m^3, which is correct. (Appropriate choice of units often means that this problem can be avoided.)

A table from a word processor or spreadsheet does not need to include many drawn lines. A horizontal line under the column headings, and possibly at the top and the bottom of the table, are all that are needed. On a handwritten table it is normally helpful to have vertical lines as well (and horizontal lines between each row if you are using plain paper). But lines should be drawn only when they make the table clearer.

If there is more than one table in a report, each table must be numbered. In a long report which is divided into numbered sections, tables in Section 3 should be numbered Table 3.1, Table 3.2 etc.

Table 6.1 Experimental results

Volume of water (measured) V (m^3)	Time (measured) t (s)	Discharge (V/t) Q (m^3/s $\times 10^{-3}$)	Head difference (measured) H (m)	$H^{1/2}$ ($m^{1/2}$)
0.100	79	1.27	0.014	0.118
0.200	71	2.82	0.070	0.265
0.200	43	4.65	0.189	0.435
0.300	41	7.32	0.469	0.685
0.300	33	9.09	0.722	0.850

Table 6.1 is an example of a table of experimental results, set out in the standard way. Tables can also contain descriptive information; Table 6.2 is an example.

Table 6.2 Descriptive information

Item	Type	Supplier	Model
Valve	Butterfly (100 mm dia)	Crane	Gem Wafer 40R series
Actuator	Pneumatic Double acting	Norbro	20RKA40
Positioner	Electro-pneumatic	IVP	K80P
Digital to analogue converter	12 bit, current output	3D	GPIS (modified)

6.3 Graphs

x–y graph

The most common type of graph for engineering students is the ordinary two-dimensional graph drawn with Cartesian (x,y) coordinates, used to investigate the relationship between two parameters. In Maths, when *y* is a function of *x*, *y* is plotted on the "vertical" axis, *x* on the "horizontal" axis. If, in an experiment, we are looking at the way parameter B is affected by changes in parameter A, B is plotted on the *y* axis, A on the *x* axis.

When *x–y* graphs are drawn by hand they are usually plotted on graph paper. Such a graph should be clear and easy to read, and should also be capable of representing the accuracy that is aimed for in the analysis. Larger

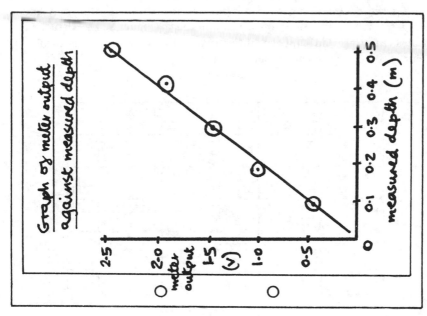

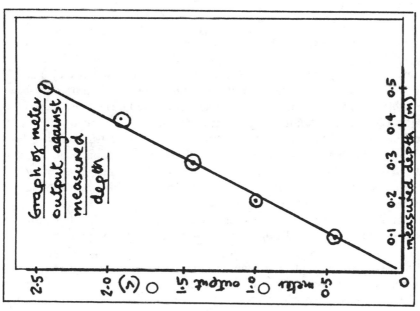

Figure 6.1 Positioning the graph on the graph paper. The graph on the left has its axes drawn along the edges of the squared part of the sheet. This has meant that the labels and title must be squeezed into small spaces. The graph on the right is better.

scales allow more accuracy, but it is not normally a good idea to draw the axes along the edges of the squared part of the sheet, as this may mean squeezing the labels and titles into a small space (Figure 6.1). Also do not use scales that are difficult to interpolate (Figure 6.2). If accuracy is sufficient, good communication becomes the priority.

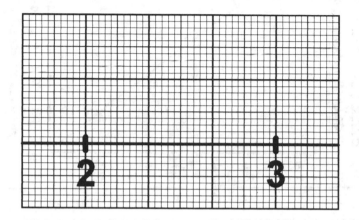

Figure 6.2   An inconvenient scale – where is 2.42?

The relationship between the scales of the x and y axes also affects the clarity of a graph. The graph on Figure 6.3(a) can be made clearer by adjusting the scales; the result is on Figure 6.3(b). It is sometimes suitable for the labels of an axis to start at a value other than zero. Figure 6.4 shows a case where this helps with clarity. The axis must be labelled very clearly to prevent misunderstanding; it is a good idea not to intersect the x and y axes, so that there is no apparent origin. (Plotting graphs in this way can also be a technique for deliberately misleading the reader, so be careful. Figure 6.5(a) reveals that the value of a company's shares has been fairly steady over a four year period. Figure 6.5(b) "shows" that the value of the same shares has increased dramatically.)

Information used to label each axis should be: description of parameter, symbol (if appropriate) and units. The graph should have a number and a title.

Many engineering graphs have measurements plotted as points, and a line which represents a best guess at the general relationship. You make an important decision when you draw this line, and you must make it carefully. Here are some questions to ask.

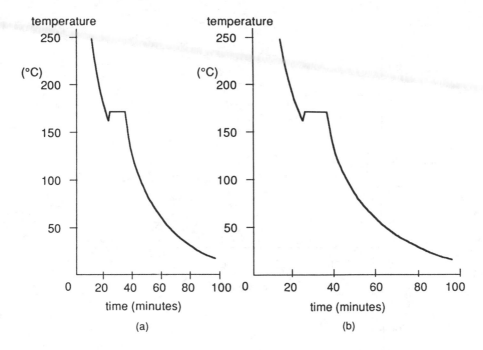

Figure 6.3 Choosing appropriate scales.
The scale of the *x*-axis on (a) is too small. The scale on (b) is better.

(a) **Should any of the points be ignored?** (You should only do so if there is a *good* reason.)
(b) **Should the line pass through the origin?**
(c) **Should anything else be considered** (for example the results of similar tests already completed)?
(d) **What type of relationship do the points suggest?**

If, after answering these questions, you decide that the line should be straight, you can fix its position by eye or by carrying out a regression calculation. If you decide that the line should be a particular type of curve, you can calculate the equation (especially if you have an advanced calculator), or you can draw a curve by eye. The benefits and dangers of allowing a computer to do these jobs are discussed in Chapter 7 (Computer aids).

If you are comparing the points with an established or theoretical relationship, there is no need to draw any line other than the one that represents that relationship (even if the points are not close to it).

If different sets of points are being plotted on one graph you can use different symbols for the different sets (Figure 6.6). If the different sets of

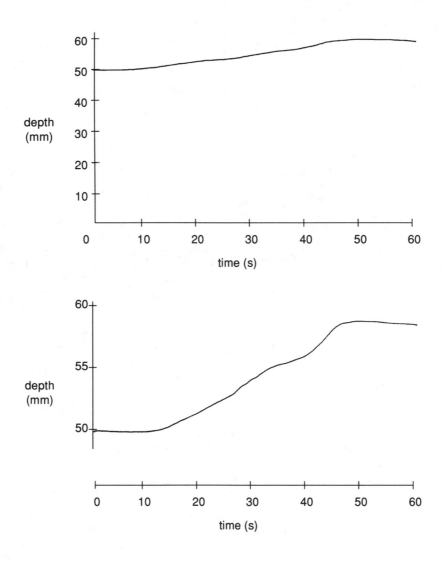

Figure 6.4 Axes do not have to start at zero
By having a larger scale, and no zero, on the y-axis (as on the lower graph), the data
on the upper graph can be plotted more clearly

points have different lines, it may be possible to identify the lines from the symbols. If not, you may need different types of line: dashed, dotted, chained etc. If the graph is to be copied on an ordinary photocopier, you cannot use colours to distinguish points or lines.

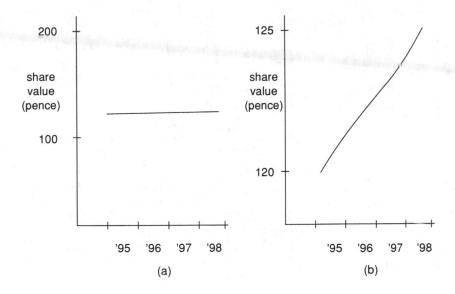

Figure 6.5 *Axes can mislead*

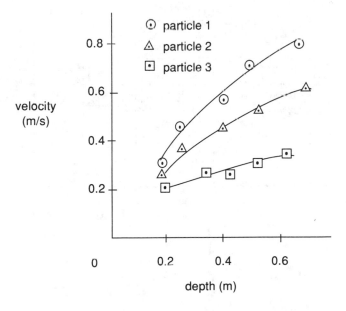

Figure 6.6 *Using different symbols for plotted points*

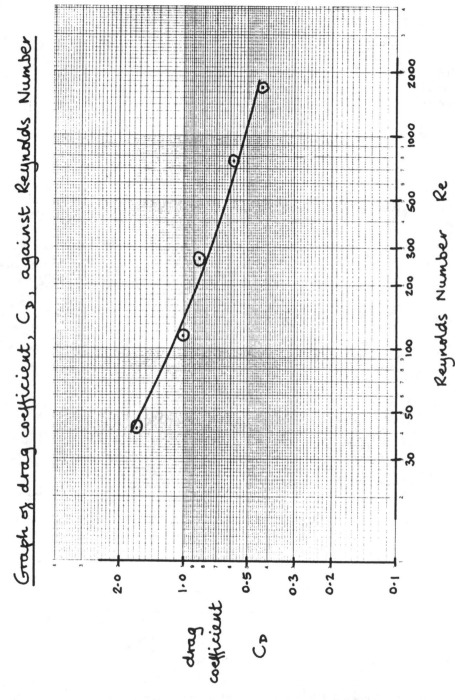

Figure 6.7 Log scales

It may be suitable for some graphs to have **log scales** rather than linear scales. You must decide how many cycles you need on both axes before you start. There is an example on Figure 6.7 (which uses three cycles on the *x* axis and two cycles on the *y* axis). On this type of paper it is normally necessary to draw the axes along the edges of the ruled part of the sheet. An alternative to log scales is plotting the log of the number on linear scales, but this does not allow values of the parameter to be read directly from the graph.

The following types of graph are also occasionally useful to engineering students.

Bar chart

Bar charts allow simple comparison of data. They are one-dimensional graphs in which the magnitude of something is represented by the length of a vertical or horizontal bar. The bar can be divided to show how something is made up; this gives a "stacked" bar chart (Figure 6.8).

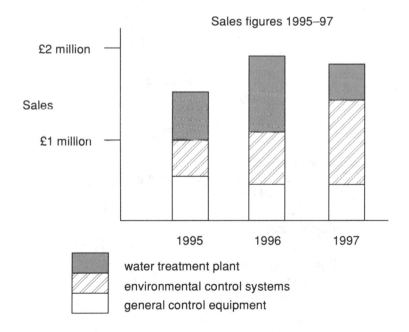

Figure 6.8 Stacked bar chart

Another type of bar chart, called a Gantt chart, shows a programme of activities. It is drawn horizontally, and the position of each bar represents the planned start and finish of an activity on a time scale (Figure 6.9).

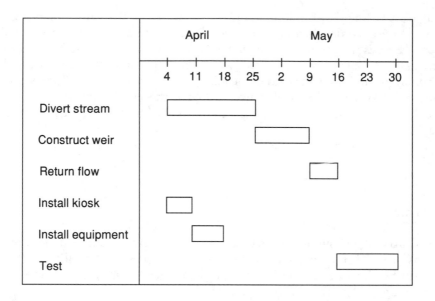

Figure 6.9 Gantt chart

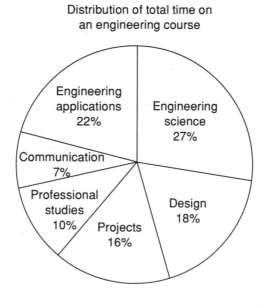

Figure 6.10 Pie chart

Pie chart

These are circles divided into segments whose areas represent relative proportions of the constituents of a whole (Figure 6.10). They communicate this information clearly and fit with popular images like "a large slice of the cake".

Histogram

A histogram can be useful for representing the general distribution of data (Figure 6.11).

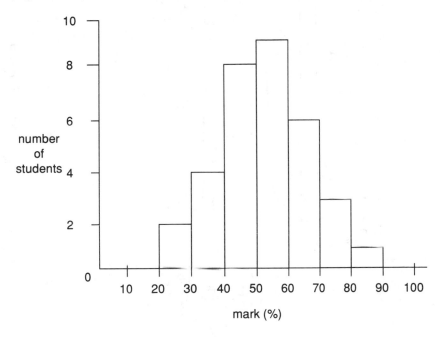

Figure 6.11 Histogram

6.4 Diagrams

A diagram is a visual representation of a physical thing, a procedure or an idea.

When engineers need a visual representation of something in full detail, they produce **engineering drawings**. These are an important form of communication, but are not within the scope of this book.

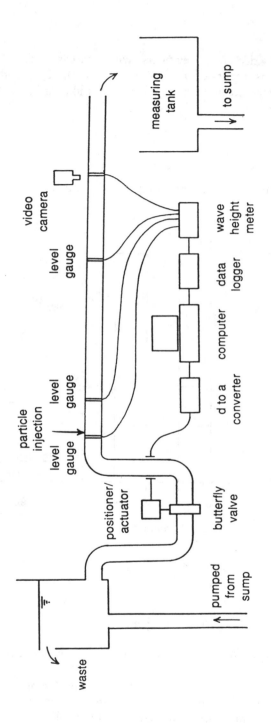

Figure 6.12 Diagram of experimental installation

A diagram nearly always has an element of simplification. Simplicity and thoroughness tend to be in competition – the more you have of one, the less you have of the other. For clarity we would like our diagram to be simple, but perhaps that means we are unable to represent important details. That is where words come in. A diagram is usually most effective when used in partnership with words. Describing something by diagrams alone can be difficult (as anyone who has tried to construct an item of furniture using only the manufacturer's diagram will appreciate). It may also be difficult to describe something by words alone. If the right diagram goes with the right words, both can be kept simple but the meaning can be clear.

Representation of physical things

A simple line diagram, in conjunction with words, can convey how a physical system operates (Figure 6.12). The diagram should show no more than is needed. If particular detail is needed, a new diagram that concentrates on that aspect should be drawn (Figure 6.13).

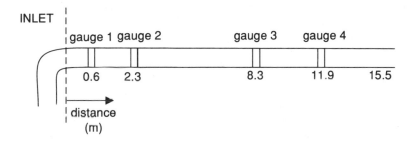

Figure 6.13 Detail of particular aspect.
Location of level gauges

Of course you will regularly see diagrams more elaborate than these in engineering text books. You may need to produce complex diagrams yourself, to show, for example, the details of a component. My point is that your diagrams should be no more detailed than they need to be to provide, in conjunction with words, a clear description.

Labels should be added where they help (as in Figure 6.12). You should include lines pointing from the labels to the details on the diagram if they are needed, but they rarely are. If a diagram is drawn to scale, the scale should be given. On a site plan or map, the direction North should be indicated.

One mistake when drawing diagrams is to try too hard to represent the appearance of something in addition to the information you are really trying to convey (how it works, or fits together, for example). Be clear about the purpose of each diagram. If you do wish to communicate the appearance of something which exists and to which you have access, you could use a **photograph**. A good photograph can save a writer a lot of work. (But photos may not be good at showing how things work or fit together.)

Representation of procedures

When we considered the writing process in Chapter 5, we summarised the procedure on a diagram (Figure 5.3, p 45). This sort of diagram, with obvious symbols and an informal style, can communicate information simply. Diagrams representing procedures can use more formal structures and systems of symbols, but they all have one characteristic in common: the reader's eye is drawn through the diagram along defined pathways.

A simple branching diagram can demonstrate relationships or links (Figure 6.14), or organisational structures.

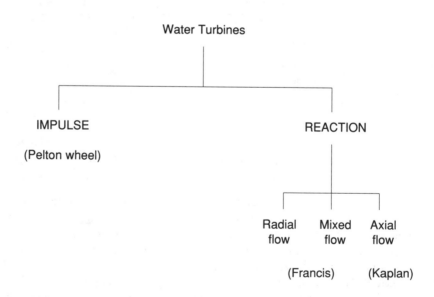

Figure 6.14 Simple branching diagram

A flow chart can show a sequence of processes or decisions. The symbols used in the example on Figure 6.15 are standard, and are defined on the diagram.

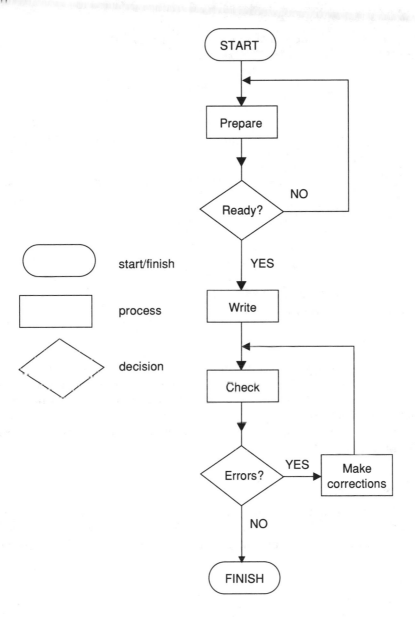

Figure 6.15 Flow chart

The collection of activities which make up a project can be represented on a precedence network (Figure 6.16).

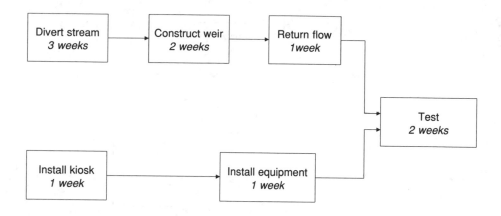

Figure 6.16   Precedence network

Representation of ideas

Design ideas

Preliminary design ideas are often best represented in the form of sketches (Figure 6.17). The effective use of sketches to develop and communicate design ideas is a valuable skill for engineers. In an interim design report the sketches would be supplemented by description and explanation in words. If the design is taken to completion, the ideas will end up as engineering drawings.

Analytical ideas

Representation of analytical ideas is often by a combination of graph and diagram. Figure 6.18 is a "pressure diagram", showing liquid pressure increasing linearly with depth, and indicating the position of the resultant force.

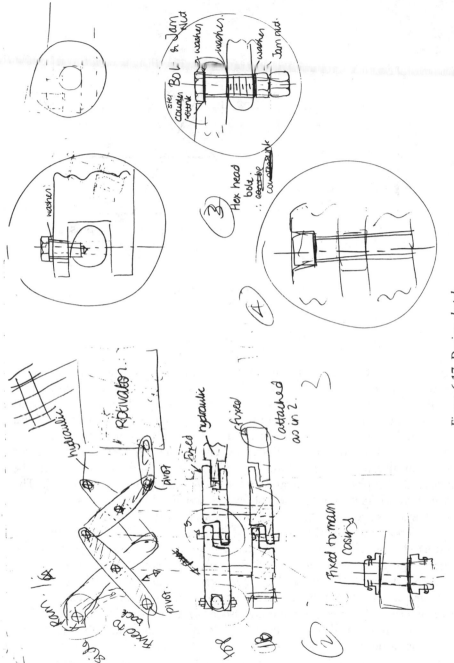

Figure 6.17 Design sketch

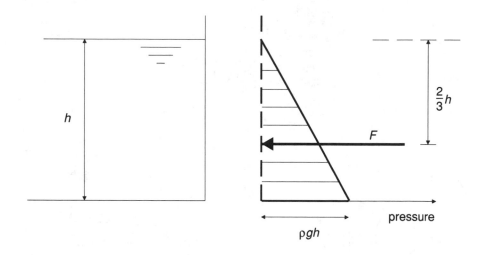

Figure 6.18 Pressure diagram

6.5 . . . and words

Where information is being presented in the form of tables, graphs or diagrams, words provide the links. They also convey the information that is best left in words. The text acts as a sort of host – introducing tables, graphs and diagrams when they are required. It follows therefore that all tables, graphs and diagrams must be referred to. If a table is not referred to from the text, it is likely to be missed by the reader, however beautiful or important it may be. Therefore all tables, graphs and diagrams must be numbered. Graphs and diagrams are usually numbered as "figures". The word **Figure** often seems to be abbreviated to **Fig.**; I can't see why – only two characters are saved, and it's quite a short word anyway.

In a report with numbered sections, the diagrams in Section 2 should be numbered Figure 2.1, Figure 2.2 etc. We should not call this a "decimal system" since the order goes . . 2.9, 2.10, 2.11 etc; 2.12 does not come between 2.1 and 2.2, as it would if these were real decimal numbers.

Figure 2.7, Table 5.1 etc. should start with capital letters because they are titles. Every table and figure should also have a title or descriptive caption, printed and positioned using a consistent format. A table usually has its title at the top; a figure usually has its title at the bottom.

In a report, the best position for a table, graph or diagram is as close as possible to the text which refers to it. A small diagram placed between sections of text can make a page look interesting and attractive, but it should be placed at a break between paragraphs. If ten pages of graphs are referred to at one point, it would be better to place them in an appendix (more about this in Chapter 10, Reports).

The results of a simple laboratory experiment are commonly given in a table and then plotted on a graph. However where there is a great deal of data you should not present it all in both tables and graphs.

Checklist for a table

- is it numbered?
- does it have a title?
- is it referred to from the text?
- is it explained/discussed in the text?
- is it well presented and clear?
- does each column have an unambiguous heading? units?
- do all values have appropriate and consistent accuracy?
- are all columns necessary?

Checklist for an x–y graph

- is it numbered?
- does it have a title?
- is it referred to from the text?
- is it explained/discussed in the text?
- is it well presented and clear?
- are the parameters on the correct axes (y against x)?
- are axes labelled with parameter name and units?
- are the axes and scales drawn so that
   there is enough space for titles and labels
   interpolation of the scales is easy
   the graph is not compressed on either axis?
- is any assumed line based on sound interpretation?
- are different points and lines distinguishable?

Checklist for a diagram

- is it numbered?
- does it have a title?
- is it referred to from the text?
- is it explained/discussed in the text?
- is it well presented and clear?
- does it show what is needed and no more?
- are scale, labels and key given if needed?

Further reading

Huff, Darrell *How to lie with statistics*. Penguin, 1991. This book offers entertaining and instructive light reading on the presentation of data and statistical parameters. It is more about communication than mathematics. It's a classic, first published in 1954. The examples are a bit out of date, but the message is timeless.

Pentz, Mike and Milo Shott (ed. Francis Aprahamian) *Handling experimental data*. Open University Press, 1988. This is a useful book. The main topics are accuracy, errors, graphs, statistics and units.

Stanton, Nicki *Communication*. Macmillan, 1990. The author gives a thorough treatment of tables, graphs and diagrams in non-technical applications (in her Chapter 18).

To learn more about technical aspects of engineering diagrams, refer to engineering text books (rather than books on communication).

7. Computer aids

This chapter considers ways in which computers can help engineering students to communicate, in particular to produce reports. It does not examine the merits of specific software packages, or explain how to operate them. The aim is to consider the potential of computer aids, and give general advice about their use.

7.1 Word processing

The computer aid to communication most commonly used by engineering students is the word processor. All engineering students have access to word processing facilities. If you are not already using a package, listen hard and learn quickly when you are introduced to one early in your course.

If you have a choice of packages, or are choosing one to use at home, consider your requirements carefully. Engineering students need mathematical symbols, Greek letters, subscripts and superscripts, and a facility for producing equations.

Become very familiar with the chosen package before you need to use it for a major report. Aim to word-process as many of your written assignments as you can, even when that is not required.

Part of the writing process

With a word processor, do you still need pen and paper? The answer for most people is yes. The writing process (Chapter 5) begins with ideas, allowed to flow out in an unstructured way. If it suits you, you could let these ideas flow through the keyboard, and later work at sorting them out on the screen. But most people prefer scribbling ideas on paper – it seems to allow a freer state of mind, and can be done anywhere, any time. Whatever medium you use for listing your ideas, you should still give full emphasis to the ideas–sorting–structuring stages of the writing process. Staring at a blank screen is just as pointless as staring at a blank sheet of paper.

When you start turning ideas into text, you may go straight to the word processor, or you may prefer to continue working by hand. There is certainly no need to aim for perfection in any hand-written draft as you will shortly have the opportunity of improving it as you type it into the computer (and any number of opportunities for revising it subsequently).

You must integrate the word processor into your own version of the writing process in the way that suits you best.

Typing and formatting

Be careful with punctuation. You should leave one space after a comma (or colon or semicolon), and two spaces after a full stop. Do not leave a space before a punctuation mark. At the end of a paragraph either go on to a new line and indent ("tab" in), or leave a blank line and don't indent. In reports and letters it is more common to leave a line (without indenting) in order to create a more definite break; in books it is more common to indent (without leaving a blank line).

To emphasise words you can make them **bold,** <u>underline</u> them, or use CAPITALS or *italics* (or a combination). Being able to emphasise words or sections in this way is useful, but it follows that you should use these options only for giving emphasis. A whole document in italics or bold usually looks like a mistake.

An important use of emphasis is in headings. It is common to make main headings bold and in capitals, and subheadings just bold:

1. INTRODUCTION
1.1 Aim of study

The main thing is to be consistent – always use the same method of emphasising subheadings, quotations and so on.

You may also be able to choose from a wide range of type sizes and fonts (designs of typeface). The type size will have an effect on the number of pages needed for documents, and on the ease with which the text can be read. The choice of font (Courier, Times etc) is a matter of taste. You must experiment and make a decision. Be cautious about choosing an unusual font. You (or more importantly your reader) may get tired of it after a few pages.

You will have control over other aspects of the printed page which can be used to make your work look professional. You will be able to set the margins (remember that space may be taken by the binding). You can decide whether or not the page should be fully justified. If it is, both left and right margins are vertical lines, with the space between the characters on each line automatically adjusted. The main alternative is for the page to be left justified only, in which case the right margin is ragged and the space between the characters is constant. You must decide which you prefer.

If your report is longer than a few pages, use the facility for numbering pages.

If your report has a title page, make it clear and impressive by centring the text, making it bold etc. There is an example on Figure 7.1. (If your course has a standard cover, or format, you must of course use that.)

Checking facilities

Remember that after you spell-check a document it is still likely to contain spelling mistakes, as in

Hear arc sum mistakes.

This could arise either because you confused two words and picked the wrong one, or because you made a typing error which produced, by accident, the correct spelling of a completely different word.

More powerful checking options that include grammar, sentence construction and use of words, can pick up some of these mistakes, but they won't spot everything.

So, using the computer to check your text is a very good idea, but you must also check the document carefully yourself. Don't let the professional print quality and the computer-based checking facilities lull you into a false sense of security.

Precautions

Here are two final points about word processing.

1. Remember to **save** regularly so you can't lose a lot of work in a disaster like a power cut or network failure; also you should **backup** on to another disk so that you have more than one saved version.

2. Remember trees. Word processing, by encouraging endless revisions, offers great opportunities for wasting paper. At least try to use both sides of the paper by printing the newest draft on the back of an old one. And read your work carefully on the screen before printing out.

University of Devon

School of Engineering

BEng (Hons) Manufacturing Systems Engineering

FINAL YEAR PROJECT

Title: **Modelling of manufacturing efficiency**

Student: **Angela Fulchrum**

June 1996

Figure 7.1 Title Page

7.2 Other computer aids

Spreadsheets

Think of a spreadsheet as the equivalent of a word processor for numbers. You can start off feeding in rough ideas and fragments of data. You can add more data, revise it, format it, perform calculations, plot graphs and produce beautiful printed tables with titles, headings and comments. Spreadsheets can be very useful to engineering students.

Graph plotting

Many spreadsheet packages have good graph plotting facilities. There are also packages primarily designed for producing graphs, and many of these include some spreadsheet functions.

Used well, graph plotting software can produce attractive and clear graphs. Used badly it can do more harm than good. The point is that the package does not do all the work; and if you assume it can, then you are using it badly. Questions (a) to (d) in 6.3 (p 55) can only be answered by an intelligent human being. Once they have been answered, the information must be passed to the computer. Another aspect often overlooked when engineering students use graph plotting packages is the labelling of the axes. If the axes are labelled ambiguously or with an unhelpful increment, the graph is not clear.

Diagrams

Some word processing packages include line drawing facilities for preparation of diagrams; others can import graphic images prepared by computer-aided design and drafting packages. Many technical and business software packages give output in the form of diagrams.

The best quality output will come from the best printer, but you don't need an expensive printer to produce reasonable diagrams.

Computers may allow you to draw complicated diagrams with ease, but remember that simple diagrams are often clearest.

Computer aids generally

Computer aids to communication can save time and effort, but their main benefit is in improving quality. When you use computer aids you should always aim to make your communication better than it would have been without them. Yet it is easy to fail to achieve this, as in the case of a graph which looks beautiful but which cannot be read because the scales are labelled ambiguously, or a document that has been spell-checked by a computer but not a human and still contains distracting mistakes.

The use of computer-aided communication in all forms will go on increasing. The technology will enable us to communicate more quickly, more accurately, more beautifully, in more different ways. But the effectiveness of the communication will still depend on the judgement of the people who are communicating. The changes will place more, not less, emphasis on our ability to use what is available to communicate well.

8. Laboratory reports

The function of a laboratory class is to allow you to learn more about a subject and to develop skills in practical investigation. One of the most important of these skills is communicating results.

Some lecturers insist on a rigid format for the presentation of lab reports. Others may simply want to see results and analysis. Obviously it is important to know what is required and to try to produce it. As with all other forms of engineering communication the most important aim is to be clear.

These pages are designed to help you write clear lab reports. It is an important skill to acquire because lab reports are engineering reports in miniature. A lab report is the perfect place to start practising technical writing.

The rest of this chapter discusses the contents of a detailed report on a laboratory class. The report on a more prolonged laboratory study will have an enlarged format based on the principles of this chapter and those of Chapter 10 (Reports) and Chapter 12 (Final Year Project Reports).

If the format for your report has not been specified but you have been asked to write a "full" report, your list of headings should be something like this:

Titles
Aims
Theory
Apparatus
Procedure
Results
Analysis
Conclusions

We will look at these headings one by one, and consider the sort of information that should be included, with tips on good practice. If you have been asked to present only part of the list, concentrate on those sections.

Titles

These are usually:

> title of experiment
> name(s) of student(s)
> date of experiment

In a long report the titles can have a page to themselves; in a short report that is a waste of paper.

Aims

This is a brief statement of what you set out to determine, investigate, test or confirm.

Theory

For an engineering laboratory class this will probably involve one or two explanations, and the statement of key mathematical expressions. Unless you have been asked to do so, there is no point in writing out standard derivations that are given in the textbooks.

Your experiment may not be related to theories of engineering science; it may for example be a standard test procedure. Some preamble to what you actually did will still be important: it will describe the purpose of the test in practice, or how the results would be used. In this case the section might be entitled **Introduction** or **Background**.

Apparatus

Alternative titles are **Equipment** or **Experimental Installation**. In this section diagrams and words should be in partnership, as discussed in 6.4 (p 61). Diagrams should be as simple as possible. Their main aim will be to show how something works, not what it looks like. Clarity is more likely to be achieved by simple sections than by elaborate perspectives.

A description of instrumentation may be necessary. The section may also contain information on the design and specification of a model or test specimen.

For the diagram and description, think about the level of detail that is really necessary; you will have to stop somewhere. The general rule for full reports is that there should be enough information on Apparatus and Procedure for someone else to be able to repeat your experiment.

Procedure

This is the section in which you describe how you conducted the experiment. There is a well established tradition of using the passive voice when describing experimental procedure. In the passive voice the object of a normal sentence (active voice) becomes the subject.

Active: Mary unlocked the door.
Passive: The door was unlocked by Mary.

Many lecturers will expect you to use the passive, and there are some good reasons for it. You might say to someone, "I measured the thickness of the specimen in four places using a vernier". That is the active voice, and the subject of the sentence is "I". Yet that subject – the person who carried out the measurement – is not important. In fact, since descriptions of experimental procedure must be precise, perhaps you ought to point out that you were working in a group of three students, and Pete took some of the readings and Laura took others. Oh no, hang on, while you were measuring the thickness of the specimen, hadn't Pete nipped out to phone his bank manager?

The identity of the person who carried out the procedure is, or should be, irrelevant, and the passive voice is a device for avoiding constantly making that person the subject of sentences. So we write

The thickness of the specimen was measured in four places using a vernier.

However, the passive voice is not the only device that can be used, and the conventional style for describing procedure can be kept very simple. For example, you might have written in your notes:

initial readings   –   room temperature
                       atmospheric pressure

The simplest way of writing that into a sentence is

The initial readings were room temperature and atmospheric pressure.

That is not the passive, but it is simple and appropriate, and still avoids the unnecessary human subject in the sentence. Clumsy passive voice expressions that are common in lab reports like "it was ensured that" and "it was observed that" can be avoided. Instead of "It was ensured that the apparatus was level by adjusting the foot-screws" you could write

The apparatus was levelled by adjusting the foot-screws.

That is still passive but not so awkward.

So, I suggest that you do not describe your procedure using a "we did that" or "I did that" style. You will probably need to use the passive voice, but not for everything. Avoid the more clumsy passive voice constructions.

Be careful with the word **then** – it can be addictive. Once you use it in the description of a procedure, it is very hard to stop using it. (This was done. Then that was done. Then another thing was done. Then . . .) Your reader will assume that you are describing the operations in sequence, so there may be no need to write **then** at all.

Finally, keep your description simple, don't write more than is needed.

Results

Refer to Chapter 6 for the best way to handle tables and graphs within a report.

If your experiment gave a reasonable number of results, it is probably best to present them in the form of a table. You must distinguish between values that were actually measured and values that were calculated from the measurements. You should include all measurements in the form in which they were recorded. Measured and calculated values do not necessarily have to be presented on separate tables; using one table may aid calculation and prevent repetition. Think carefully about accuracy; give only the appropriate number of significant figures for measured and calculated values.

Remember to place a clear title at the top of each column defining the parameter, giving the symbol (if used elsewhere in the report) and the units. Don't use a column for identical values of a constant: a column entitled "gravitational acceleration, g, m/s^2" filled with the number 9.81 is pointless and distracting. Where a column contains calculated rather than measured values, give the expression used in the calculation at the top of the column. Even if a calculation is quite involved, present it on the table, do not write out the calculations longhand. (If you feel that the calculation is too complicated to be understood from the column heading, you could write out one sample calculation in full.) You will probably want to use a computer spreadsheet for complicated tables – but the guidelines above still apply. A sample table of experimental results has been given as Table 6.1 (p 52).

Your observations may not all be measurements; some of your results may be qualitative not quantitative. If descriptive observations are important, they should also be given in the **Results** section. They too can often be presented in the form of a table.

To finish this discussion of presentation of results, here's a suggestion that is as much to do with the conduct of the practical work as the presentation of the report. Try to record measurements in the laboratory **neatly**. It helps you to plan and understand what you are doing, and can save time. If your record is tidy, it can go straight into your report, with calculations carried out on a spreadsheet. Life is short!

Analysis

This is an important part of the report, though of course the approach to the analysis will depend on the nature of the experiment.

Analysis of the relationship between parameters is likely to involve plotting graphs (see 6.3, p 52). Interpretation of the graphs, and perhaps comparison with theory, will often be the "aim" of the experiment.

Presentation and interpretation of graphs of experimental results requires care and judgement – often more judgement than a computer is capable of. That is why your judgement when plotting a line by hand through your points may be of more value than the quality of presentation given by a computer.

The way you describe the outcome of your analysis, for example the comparison between experimental results and theoretical relationship, is important. Expressions like "there is reasonable agreement" are vague, though there are worse, for example: "the law is verified because all the points are near the line" (even when they are not). The obvious problem is that words like "reasonable" and "near" are not precise. You should be as precise as you can. Give a maximum percentage difference, or a regression coefficient, or some other appropriate characterisation to suit the nature of the experiment.

One aspect you should comment on, even if your results are "good", is possible errors. You should identify elements of your data which you consider to be particularly affected by errors, and also identify types of error which may have a general effect on your results. The expressions "experimental error" and "human error" are unhelpful – again, be precise. What caused the error? How large was it likely to be? Anyone who has taken a measurement can, and should, form a judgement about the error that might be associated with it.

Conclusions

The real purpose of lab classes is to allow you to learn something, so perhaps the conclusions should start "I learnt the following". But in your report you have stated an **Aim** ("to determine . .", "to investigate . .") and the conclusions should relate to this. You should summarise the outcome of your experiment: what you have determined, or what the investigation has shown.

Checklist for a laboratory report

- is it the sort of report your lecturer wants?
- have you covered all the items expected?
  ... depending on what you have been asked to include ...
- have you described a clear aim?
- could someone else carry out your experiment on the basis of your description of apparatus and procedure?
- do your tables, graphs and diagrams comply with the guidelines in Chapter 6? (there is a separate checklist for each)
- are all values quoted with appropriate accuracy?
- is it clear which of your results are measured and which are calculated from measurements?
- have you thought about what the results really mean?
- have you drawn conclusions that relate to your aim?

Further reading

Pentz, Mike and Milo Shott (ed. Francis Aprahamian) *Handling experimental data*. Open University Press, 1988. For comments on this book, see Chapter 6, Further reading.

9. Essays and exam answers

As an engineering student you will not write as many essays as an arts student, but you will probably write some. There may be a general interest essay written for your tutor in the first year of the course, or an essay (or "seminar report") for a subject like Professional Studies or Engineer in Society. Then there are exams. Questions in Management or Economics may call for short essays. Science subjects like Materials Science or Geology will involve descriptive exam answers which have much in common with essays. After you graduate you may take exams which involve writing essays in order to qualify as a Chartered Engineer.

In this chapter we will deal first with essays (not in exams) and then with the special problems of writing in examinations.

9.1 Essays

The process of writing an essay

Practising engineers do not write essays at work. Engineers and engineering students come across essays only as forms of learning and assessment. The essays they write are generally fairly short. They are descriptive rather than mathematical, and usually include an element of discussion. Essays tend not to be divided into numbered sections, but may include subheadings to clarify structure. Essays are personal, not simply factual; so while the assessor may look for a number of specific points, there will be no single correct answer. However while your essay may include your opinions, these must be based on facts and not on flights of fancy. Essays call for knowledge, careful and balanced interpretation, breadth of thought, and good writing technique.

When writing an essay you must take care to write clear English, using the ideas of Chapters 2, 3 and 4. You should keep the style formal, and the sentences generally short.

The writing process developed in Chapter 5 will work well for essays – with a few adjustments. The outline scheme, first given on Figure 5.3, is reproduced on the left side of Figure 9.1. On the right are the equivalent

Defining	subject	title (which will be expanded)
	aim	to learn, to earn marks
	readership	lecturer who marks
Sorting out ideas	ideas	opening up title ←——→ reading
	sorting	establishing themes
	structuring	the plan – paragraph by paragraph intro – middle – conclusions
Putting ideas into words	preparing	notes for each paragraph
	writing	writing
	improving	checking

Figure 9.1 The writing process for essays

operations for essay-writing. The **Defining** stage requires no time at all. The subject is simply the essay title. That is a characteristic of essays; while the aims and scope may not be defined for you, the title usually is.

When you come to the **Ideas** stage you should think carefully about the title – all your ideas must be related to it. As you develop ideas, you are really opening up the title, and as with all writing, this is where the scope and richness of your content is determined. We will look at an example later.

The development of ideas is linked to researching the topic. It is not possible to generalise about this. If the topic requires you to evaluate arguments that you understand well already, you may want to develop your ideas to an advanced stage before researching particular examples or case studies. If your factual knowledge is insufficient to allow you to develop ideas, you will need to do some reading first.

Sorting your ideas involves establishing the themes of your essay. This may involve grouping points for and against a proposal. The result of **structuring** your ideas will be a plan for your essay. The basic unit in essay-planning is usually the paragraph. You should plan your essay paragraph by paragraph.

The rest of the writing process is as before. **Preparing** will involve making notes for each paragraph. If **writing** becomes difficult, if you are staring at a blank page, you must step back a stage. Checking and **improving** your first draft are essential.

You must be particularly careful about the way you use material from other sources. If you quote figures, you must say where they came from. Text that is quoted in full must be included in inverted commas, and all references to other people's work must be acknowledged carefully. (More information on references and quotations is given in 12.6 (p 114).)

Essay structure

When a teacher at my school said, "Beethoven's life is divided into three periods: early, middle and late", everyone laughed – the labels seemed so obvious and unhelpful. People tend to make equally obvious remarks about essays. We are told that there should be three sections: introduction, middle and conclusions. "Middle" doesn't help anyone, but the other two divisions are worth considering here.

The first paragraph (or more) should be an introduction to the topic and to your treatment of it. If you have chosen to interpret the title in a particular way, or to concentrate on one aspect, you should explain this at the start.

In your last paragraph (or more) you must give some outcome. Don't overdo it by writing something like: "I conclude that all engineers should learn Japanese and own helicopters". Just try to resolve the arguments in a realistic and helpful way. Make the reader feel that something worthwhile has been achieved.

Opening up the title

This is the key to worthwhile and imaginative essay writing. Let's take an example which requires no specialist knowledge beyond the normal experiences of an engineering student. Suppose you have been asked to write an essay with the title "**Engineering courses arc too narrow**". The title, of course, is a controversial assertion, like the motion in a debate. You are expected to consider the arguments for and against, in effect to answer the question "*Are* engineering courses too narrow?".

It's time for ideas, time to open up the title. Write the title at the top of a sheet of paper and start thinking. Write down all your thoughts in note form. Here are the thoughts you might have (written out in full).

Who decides how narrow courses should be? The answer must be: the lecturers who teach them, the profession (which approves courses via the professional institutions), and the government (which ultimately controls the length of courses).

What is the aim of an engineering course? Is it just to prepare engineers for industry, or is it to provide an education for life? Society tends to expect professional people to be "educated" as well as simply good at their jobs. In any case not all engineering graduates become engineers.

What subjects can widen an engineering course? Economics, Management, and Communication seem to be common. Some courses have the option of a foreign language. Wider? How about Business Studies, Psychology or Sociology? Wider still? Politics, Philosophy, History, Art, Literature?

Yet courses seem to be crammed full of engineering science and applications. There is scarcely time to *think*, let alone study "wider subjects". Could some of the engineering content be dropped? If so, what? Perhaps courses need to be longer.

Engineers seem to be thought of as having narrow interests and being reluctant to become involved in public life outside their profession. This may be why they lack influence and political clout. Is this connected with the narrowness of their education?

Can the experience of any student (meeting people, playing sports) be described as narrow?

So, now we've opened up the title to cover quite a wide range of interesting points. It's probably time for some researching. How different is one engineering course from another? What about the different branches of engineering? Some courses have different lengths; how is the time used? What about different countries? Have the professional institutions or the government published their views? What do your lecturers think? Off to work – there's obviously no shortage of ideas.

9.2 Writing in exams

A minority of engineering examinations call for answers in the form of essays or extended descriptions. Engineering students don't get much practice at writing in exams, and often don't do it very well. Here are some ideas that may help.

First, think about the question. Watch out for optical illusions and self-deception. Think about the question that is actually set, not the one you hoped for, or the one you would be able to answer easily. If you don't understand the question, don't answer it (assuming there is a choice, and you do understand some of the others).

Then you should begin a condensed version of "the writing process". Because of time constraints, a free-flowing ideas session is not really possible. This means you must work out the general themes or topics before you think of detailed ideas. Then as you note down the details they can form part of a ready-made structure. But **make notes** – that is the important thing. When you have completed this stage, write out your answer. Keep your style formal and your sentences short. Don't try to impress. Be factual and specific. Use diagrams and examples wherever they may help to make your meaning clear. Don't try to make jokes – the only thing that will cheer up the examiner is a good answer. When you have finished, no matter how short of time you may be, check what you have written thoroughly.

Exam technique

Here is some brief advice on exam technique in general. It applies to any type of exam. These are the three principles that I think are the most important.

1. *Don't overdo it close to the exam.* Like an athlete preparing for an important event, do some gentle work the evening before and then relax (this means that your revision must have been well planned). Don't join the overexcited group of your fellow students outside the exam room asking each other questions like "What's the definition of ...", "How do you work out ...?" – you will only waste mental energy and make yourself anxious.

2. *Read the paper carefully at the start of the exam.* Think hard about your choice of questions. Start with the ones you think are easiest.

3. *Plan your time, and monitor your progress, carefully.* Allow a generous amount of time for reading the paper at the start and for checking at the end. If the mark allocation for a question is given, use it to proportion your time. Don't allow much over-run for a question – if there's time at the end you can go back.

10. Reports

This chapter considers the general skills that are needed when engineering students write reports. Some types of report have already been introduced: lab reports in Chapter 8, and seminar reports, covered under the general heading of Essays, in Chapter 9. Other types of technical documentation, not usually referred to as reports, are considered in Chapter 11. The final year project report, often the most ambitious report that an engineering student writes, is covered in Chapter 12.

Considering all branches of engineering, the variety of reports written by students is wide. The properties of specific types of report are considered later in this chapter. However, the most important report-writing skills are those that are applicable to all types of report, and the main aim of this chapter is to help you acquire them.

10.1 Defining the task

A report is a formally structured collection of written information that a person or group wishes to communicate. A report is self-contained, is about something specific, and is written for a particular reason.

The important first stage in producing a report is **defining the task**. We considered this stage in Chapter 5 (The Writing Process). It involves

defining the subject
defining your aim in writing about it
defining your readership.

There is an element of artificiality when a student makes these definitions. The **subject** should have been set at the start of the assignment, but you may need to refine its definition. For a student there are two types of **aim**. The first is to learn and to earn marks. The second is the sort of aim that the report would have if it were being written in engineering practice. In the case of a feasibility study report, for example, it would be to assess the appropriateness of a project. You must attempt to achieve both types of aim in the way you write the report: the second explicitly, and the first implicitly. The **readership** can also be defined in two ways: either as the assessors of the work or as the readers of such a report in practice. The report must be designed for the second readership: a feasibility study report for the client. (The first readership is addressed implicitly with every attempt to achieve quality.)

10.2 Structure

A formal and tight structure is one of the particular characteristics of a report. A novel, essay or autobiography might also be carefully structured, but the structure would be woven into the writing and not necessarily obvious to the reader. The structure of a report is made explicit: the sections are marked out by headings and subheadings, and are numbered in a systematic way.

Adaptation of twin-wire wave gauge for use in pipes

1. Introduction

2. Twin-wire wave gauges
    2.1 General principles
    2.2 Traditional applications
    2.3 Calibration

3. Adaptation to pipes
    3.1 Design of gauge
        3.1.1 Requirements
        3.1.2 Selection from alternatives
        3.1.3 Installation details
    3.2 Data logging system

4. Practical tests
    4.1 Laboratory equipment
    4.2 Effect of gauges on flow
    4.3 Calibration
        4.3.1 General relationship
        4.3.2 Temperature effects
        4.3.3 Calibration procedures

5. Conclusions and recommendations
    5.1 General conclusions
    5.2 Recommended applications
    5.3 Recommended calibration procedure

Figure 10.1 Numbered headings and subheadings

There is an example list of numbered headings and subheadings for the main content of a report on Figure 10.1.

This type of numbering system is generally the most suitable for a student report of any length. The headings of the main sections are numbered 1., 2., 3. Where the main sections are split, each subsection has a subheading numbered 3.1, 3.2. These subsections may themselves be split by using minor subheadings numbered 3.1.1, 3.1.2.

Some reports have a fourth layer (3.1.2.1). I think this makes the structure look complicated, and my advice is to avoid the need for such a subdivision by changing the overall structure.

This method of setting out structure should be incorporated into the writing process (Chapter 5) at the **structuring** stage. This means that when you have sorted the ideas for your report into themes, you should establish headings, subheadings and a numbering system. This will help you to establish the clear formal structure that your report will require.

Now try out your skills at report structuring with two tests.

Test 10a

You are a "Party Consultant" (social parties not political ones). You charge a fee for giving advice on organisation of parties – anything from small gatherings to major events. It is a competitive field, where reputation is all-important. Your reputation depends on the quality of your reports.

You are to prepare a report on the party potential of the place where *you* currently live. Plan the structure of the report, including numbered headings and subheadings together with summaries of the content of each section.

Follow the ideas–sorting–structuring sequence described in Chapter 5. Don't write the report itself.

(You are not being asked to plan a specific party, but to assess the potential for parties. That means that by considering size of rooms, closeness of neighbours, kitchen facilities, parking, public transport etc, you should determine the most suitable type(s) of party for your place.)

Suggested answer, p 160.

Test 10b

Imagine that you are a final year engineering student and you have been appointed as a representative of your year. You have been asked to write a report on a meeting of the year held to receive comments on the course as a whole (let's assume it's a three year course). The comments are listed below in note form and in random order, coded **a.** to **u.**

Create the most suitable structure for the report. Write down numbered headings and subheadings, and the code letters (**a.**, **b.** etc) of the comments you would place in each section (in order). Add any other elements that you consider appropriate. Don't actually write the report.

a.   course generally well organised and what students expected
b.   2nd year design competition enjoyable
c.   1st year Maths starts too quickly
d.   Lecture Room D cold in winter and hard to see the board
e.   lecturers approachable for personal advice
f.   computing floor overcrowded
g.   paper for printers in computing floor always running out
h.   not enough feedback on 2nd year health and safety reports
i.   generally subjects are practical and relevant
j.   lecturers generally seem up-to-date in their fields
k.   computing floor – need more brief user guides for packages
l.   should be more industrial visits
m.   laboratories well equipped
n.   students should have access to laboratories in the evening
o.   librarians helpful
p.   library: should have longer opening hours at weekend
q.   hard to contact lecturers during Christmas vacation
r.   too much coursework just before exams in 1st year
s.   not enough vegetarian food in canteen
t.   final year: need classes on report writing during projects
u.   new coffee machine much better than old one

Suggested answer, p 161.

10.3 Beginning and end

We have considered how to organise the main content of the report, now we must think about some important elements that you should place at the beginning and at the end. Unless the report is only a few pages long, it should have some or all of the following.

Title page
Summary
Contents page

1. Introduction
    .
    .
    . numbered sections
    . of the report
    .
    .

    Conclusions

References
Appendices

It is often appropriate for several other types of list to follow the contents page. These could include:

List of Figures
List of Tables
List of Symbols
Glossary (list of definitions in alphabetical order)

Here is more detail on the main sections at the beginning and end of a report.

Title page

This should give:

name of institution and course
the type of assignment
the title
name(s)
date

Summary

This should be a summary of the whole report – from introduction to conclusions. A good summary is of great value to the reader. Some students are reluctant to include their conclusions in the summary. They feel that something so important should only be read in full, and then only after the main content of the report has been read. But you're not writing a detective story! You don't have to keep the outcome a secret until the last page. The convenience of the reader must come first.

The summary of a normal report should be less than one page in length. Advice on how to write a summary is given shortly (in 10.4, p 96).

Contents

The Contents page sets out the structure of the report and how one part relates to another – it does more than just show where to find things. However, showing where to find things is important too, so you must remember to number the pages of the report. Pages before the main report begins, those containing the summary, list of figures etc., can be numbered with Roman numerals.

An example Contents page for a report by an engineering student is given as Figure 10.2.

Introduction

This is likely to be the first numbered section of the report. It will include information on the topic and on the aim and content of the study. The reader must quickly learn what the report is about.

Conclusions

An engineering report in industry is likely to have the conclusions near the front, since they will be the main interest of many readers. All parts of a college report will be of interest to the assessors, so the conclusions can usually go at the end (the most logical place for a section with that name). These sorts of differences are discussed further in Chapter 18 (Professional Communication).

The conclusions should relate to the aim of the report. When the outcome is a positive proposal or selection of an alternative, the heading should be **Conclusions and Recommendations**.

References

When other publications are referred to in the text, their details must be given in a list at the end of the report. References are of most importance in

CONTENTS

Figure 10.2 Example contents page

academic reports and so they are treated in detail in Chapter 12 (Final Year Project Reports, 12.6, p 114).

Appendices

If detailed information on a particular aspect would break up the flow of text in the main report, it can be placed in an appendix. Examples are tables of results which have already been summarised in the report, or a series of graphs of which a representative sample has already been presented and discussed.

Checklist for format of a major report

- is there a clear structure, with numbered headings and subheadings?
- have you included: title page
  summary
  contents page
  if appropriate:
  list of figures
  list of tables
  list of symbols
  glossary
  references
  appendices?

- does your summary cover the whole report from introduction to conclusions?
- are the pages numbered?

10.4 Summaries

It is important to produce a good summary. Being able to summarise well is also a useful skill for other reasons: it can help you record and understand information that you have read, it can help with revision for exams, and it is a good writing exercise. Other words for **summary** are **synopsis**, which means the same, and **abstract**, which is a summary used outside the document.

I will keep my suggestions for producing summaries simple, and let you develop your own technique. Here are the main stages.

1. Read through the piece. (If it is a long report that you have just finished writing yourself this may not be necessary.)
2. Note key words and phrases.
3. Work these up into continuous prose.
4. Revise to make sure the summary is clear and comprehensive (and has the correct number of words if specified).

Lifting whole sentences from the original is not a good idea as it wastes words and means that a thorough reader of your report will read the same sentence twice.

Test 10c

Summarise this in 100 words. It is based on the Introduction to this book.

The word "engineer" can be traced back to the Latin *ingenium* meaning cleverness, or natural ability. The main business of professional engineers is to be ingenious: to come up with good ideas and make them work in practice. No engineer works in complete isolation; there is no point in having a good idea if an engineer is not able to communicate it. Poor communication can create ambiguity, even cause disasters. At the very least it gives a bad impression: a person who communicates badly will not be fully trusted as an engineer. There is a lot at stake. The quality of each engineer's achievements, the benefit to society of engineering projects in general, the status and reputation of the whole profession, all these things depend on good communication.

Yet graduate engineers are notoriously poor at communicating. There are consistent indications from employers that standards of communication among engineering graduates give rise to serious disappointment. While there is debate over the level of emphasis that should be given in engineering courses to some non-technical subjects, Management for example, there appears to be a consensus over the importance that should be attached to Communication. The common view seems to be that standards are falling, and while universities may not be to blame for the worsening standards, they can clearly play a part in reversing the trend.

Good English is not based on fixed and absolute rules in the way that Mathematics is. But like *engineering*, good writing requires judgement, and good judgement requires confidence. Engineers and engineering students should not lack confidence in communication. Good engineering and good communication have a lot in common: they both require application of knowledge to achieve results, and both call for imagination and pragmatism.

continued

Test 10c continued

Engineers and students need to follow one elementary rule in communication: be clear.When engineers write or speak they should have something specific (usually factual and precise) to say. It should be possible to separate *what* is being communicated from *how* it is being communicated. If engineers are not sure precisely what they want to say they must stop and work it out first. When they are sure, they must know how to communicate it successfully: for the receiver to understand without loss of precision.

Another characteristic of communication by engineers is that words are by no means the only medium. Numbers, tables, mathematical expressions, graphs, diagrams and drawings can all enhance communication. Words are only used where they are needed.

(about 400 words)

My attempt is on p 161.

10.5 Style

The early chapters of this book are designed to help you write good English and use a style which is appropriate for an engineering report. If you want to sharpen up your skills you should refer back to the relevant chapters:

2. WORDS spelling, use, meaning
3. SENTENCES sentence formation, punctuation
4. GRAMMAR AND STYLE

Checklist for text of a report

- have you checked spelling by eye as well as by computer?
- have you checked your use of words in all cases where you were uncertain?
- do all your sentences form complete statements?
- have you checked your use of punctuation in all cases where you were uncertain?
- have you checked your grammar in all cases where you were uncertain?
- is your style precise, brief and simple?
- does it sound right?
- is it appropriately formal?

If your English is good and your meaning is clear, you are well on the way to producing a good report.

Don't underestimate the time and effort needed to write well. Some engineers think they are not good at writing when really they have just never taken the time to try.

10.6 Appearance

Chapter 7 contains information on improving the appearance of a report using a word processor and other computer aids. After achieving the appearance you want on the computer, you should use the best quality of printer that is available for the final copy (but check the draft copy thoroughly first).

Some experts on report writing insist that the pages in a report should be single-sided not double-sided. If your report is fairly short and you don't need many copies, this may be the best way. You may in any case be submitting the original single-sided printed sheets. However if the report is long and you need a number of copies, it is much cheaper and more environmentally sensible to make double-sided copies. You should have no reservations about doing so. You should, however, try to ensure that even-numbered pages are on the left and odd-numbered pages are on the right when the report is opened (like this book).

Covers and binding

These important details depend on the length of the report and on the standard practice on your course. I suggest the following.

1-2 pages For a brief coursework assignment, barely a "report", a separate title page is not needed. Place the title information at the top of the first page. Staple the pages together and place them in a clear plastic folder to keep them clean.

3-9 pages A separate title page at the front is appropriate. Place clear plastic sheets at the front and back, and bind with a plastic slide-binder.

10+ pages Bind with a ring comb binder or better, with card covers at the front and back. If there is a standard design of cover for the course (or for the particular project) you must use that.

Appearance and first impressions are important.

10.7 Types of report

Let us now consider, in a little more detail, some of the types of reports that engineering students write (excluding those that are covered in other chapters).

Every type of report cannot be covered in detail. Writing specialised reports is a specialised business and needs the advice of a specialist. Also there is no single correct way of writing anything. Every report-writing task is different and every writer has a different approach.

Visit reports

Engineering students may be required to write reports on visits they make, for example to construction sites, manufacturing plants, processing installations, exhibitions or museums.

Planning the report will involve sorting out the information you have acquired into a clear structure. You should provide introductory information which assumes that your reader knows nothing about the arrangements for your visit. It may be clearest if you begin with a series of titles similar to those below.

VISIT REPORT

Location:
Purpose of visit:
Date of visit:
Those involved:

The introduction should go on to describe what was being constructed/processed/exhibited, with a summary of what you saw.

Some of the report can be in a narrative style.

We were shown round by Mr Graham Brown, who is the Project Engineer for the consultants, D G Gammerson and Partners. He showed us a number of interesting construction details within the Control Building . . .

However, most of the report should describe what you saw, rather than how you saw it. So rather than use a narrative style throughout:

"Then we were shown the air-conditioning ducts which were . . . Mr Brown explained that during installation there had been a problem with . . ." simply write:

The air-conditioning ducts were . . . During installation there had been a problem with . . .

Sometimes visits come early in a course, before you have learnt much about practical engineering; however you should still try to master basic terminology. Specialist dictionaries and text books can help here. You will not need to explain basic terms in your report. You should not, for example, give a detailed explanation of every common engineering material that you saw, only the more novel ones.

The purpose of your visit may be to carry out a visual inspection, or assess the condition of a structure or installation. As with any technical description, you should use clear simple language with appropriate reference to diagrams, layouts and photographs. When you carry out an assessment, you should be careful to separate factual and indisputable observations from your own judgements and recommendations.

Fieldwork reports

In many respects fieldwork reports are similar to visit reports.

The importance of good note-taking is even greater. It is best to make your notes while in the field in a hard-covered notebook, and to look after it with care. You should try to train yourself in the discipline of keeping tidy, accurate and usable records. The records may be visual, in the form of sketches or photographs. These should be accompanied by notes giving the location, direction and vertical angle of the view. If the sketch or photo is presented in the report, the same information should be reproduced, with location and direction indicated on a plan.

You may come back from a field trip with a mass of information, and the sorting out and structuring of the ideas will need care and thought. You must choose the structure which makes the report most clear and readable; it is not necessarily best to describe things in the order in which they were visited. As with a visit report, you should not assume that the reader knows the circumstances of the trip. Location, date and purpose must be made clear in the introduction.

Visual records are important. Photographs are of great value, but a diagram drawn by you may be more useful than a photograph. A diagram may not **show** more than a photo, but it may **communicate** more.

Training reports

As a student you may need to write a training report on a period of industrial experience. As a graduate engineer you may write training reports as part of the requirements for achieving a professional qualification. Professional training reports may have a defined format, with a specified method for classifying different types of experience.

A training report will include a technical description of the projects with which you have been involved, including any special features or problems. But the emphasis must be on your own experiences. You should describe your role in the project, any special responsibilities you took on, how your role developed, any problems you encountered and how you solved them.

Different periods of experience or changes in duties or location can be shown on a Gantt chart (described under Bar chart in 6.3, p 59).

An excellent way of preparing to write a training report is to keep a diary. You may even be required to submit your diary. In any case it is good practice to keep a diary as a professional record.

Test/Investigation Reports

Tests and investigations include a broad range of practical technical studies, including tests in a laboratory or on site, and investigations in the field or by computer. A particular type of test report, the student laboratory report, has been considered in Chapter 8.

A test or investigation report is likely to include many of the techniques for communicating technical information covered in Chapter 6. The general structure is likely to be:

Introduction
Description
Results
Analysis
Conclusions

Each component should be clearly separated. For example, you should present all the results before starting the analysis, even if testing was carried out in more than one stage. Results should be set out in the clearest and most logical order, not necessarily the order in which they were obtained.

For a major test/investigation which forms a final year student project, advice on preparation of the report is given in Chapter 12.

Feasibility Study Reports

In everyday speech, the word "feasible" is used to mean something similar to "possible". However the purpose of an engineering feasibility study is not to find out if something is possible, but to find out if it is **appropriate**. A number of criteria will be used, and these vary with the branch of engineering, but they could be technical, economic, financial, social and environmental. A feasibility study carried out internally within a firm may be particularly concerned with profit, utilisation of resources and consistency with the firm's goals. A feasibility study for a client may be concerned with value for money and environmental impact.

A typical feasibility study examines a problem or a need, proposes alternative schemes to solve or satisfy it, compares the alternatives using the full range of selected criteria, and makes recommendations.

A feasibility study report is quite different from a proposal (to be covered in the next chapter). In a proposal the writer sets out from the start to recommend a particular course of action. In a feasibility study the engineer makes objective judgements about the options.

Design reports

Design reports may be needed at various stages in the progress of a design. A report before the start of design will specify requirements. Depending on the branch of engineering, this could be called a design specification or a brief. It must be thorough and precise, as it will have a crucial influence over the development of design. Specifications are considered in the next chapter.

In the early development of the design, reports may be needed to communicate the ideas and concepts being considered. In many branches of engineering the ideas are best described visually, with sketches and diagrams forming the basis of the communication. (An example of a design sketch has been given as Figure 6.17 (p 67).)

When the design is complete it will be communicated by engineering drawings and defining documents such as a specification. The design will be based on detailed calculations which must exist in a clear, readily understood, and easily checked form. Engineering students are likely to be required to submit the complete design with all drawings, documents and calculations, together with an introduction explaining the requirements, describing the development of the concepts, justifying the choice of particular solutions, and setting out the approach to the design calculations.

Progress reports

Communication plays an important part in maintaining progress on a project. Progress reports may be quite simple, giving comparisons between actual achievement and the planned programme, and information on resources and costs. For a project of any complexity, the analysis can be carried out by a project management computer package. Progress and future programme are shown most simply on a Gantt chart, and the logical relationship between activities can be shown on a precedence network (both introduced in 6.4, and shown on Figures 6.9 and 6.16). The package will produce these and other diagrams showing the programme, utilisation of resources, and costs.

A progress report should contain the appropriate level of detail. For members of the project team the precise scheduling of each activity is of interest. For a more senior manager or the client, less detail is appropriate: an overall idea of progress and costs is sufficient.

Programmes can go out of date quickly in engineering, so progress reports must be circulated regularly and frequently.

Checklist for content of a report

- does the report have clearly stated aims?
- have they been achieved?
- is the reader quickly told what the report is about?
- is the material clear, accurate and well presented?
- have you taken appropriate opportunities to make recommendations?
- is the report worth reading?

Further reading

The following three American books contain good advice. They are all written for students, though they deal mostly with professional forms of communication. They include guidance on writing the different types of report covered at the end of this chapter. They are all thorough and include plenty of examples, but they are not brief. (Roze is the shortest at about 300 pages.)

Pauley, Steven E. and Daniel G. Riordan *Technical report writing today*, 4th edn. Houghton Mifflin, 1990.

Eisenberg, Anne *Effective technical communication*, 2nd edn. McGraw-Hill, 1992. This is particularly thorough on use of English.

Roze, Maris *Technical communication – the practical craft*, 2nd edn. Merrill-Macmillan, 1994.

11. Proposals, specifications and manuals

Engineers and students write some important technical documents that are not called reports. The significance of this type of writing varies from one branch of engineering to another. Engineers write the documents to satisfy specific needs within the industry; students write them as a preparation for their careers. For students, there may be a need for artificial realism – a member of staff playing the part of the client, for example.

This chapter can only contain general statements about these types of document, as each specific application is different. You should refer to specialised texts for more information: your lecturers are likely to make recommendations, and there are some suggestions in **Further reading** at the end of the chapter.

Many of the skills needed for report writing, covered in the previous chapter, apply to these documents. Structuring will be important, and you are almost certain to use numbered sections. Other aspects, including summary writing, arranging contents and using diagrams, also follow the same principles as reports.

11.1 Proposals

Proposals are documents designed to persuade someone to accept an idea. Their aim is to win work for your team or your firm. Format and length vary widely depending on the type and size of the proposal. You may be asked to follow a particular format by the potential client or funding body; part of the proposal may even go on a standard application form.

The fundamental characteristic of a proposal is that it consists of two contrasting types of document subtly combined into one. On one hand it is a cold and factual technical document, giving facts and figures which lead to an apparently objective statement of benefits. On the other hand it tries, as surely as any advertisement or sales brochure, to persuade the reader to accept ideas. Yet it must not sound like an advertisement; it must sound like an objective statement. And the facts must be accurate, not misleading.

The elements of a proposal are likely to be as follows.

Problem definition

This will only refer to problems that your proposal will solve.

Objectives

Only those that your proposal can achieve in solving the problem.

Proposed solution

This is a summary of the work you propose.

Benefits

The benefits should be specific and quantifiable. You should show how they would be measured.

Programme and resources

The programme shows when stages of the project will be completed, and also demonstrates that you have thought carefully about the practicalities of the work. A Gantt chart (described under Bar chart in 6.3, p 59) can be used to illustrate the programme and also to provide a visual summary of the stages of the project. A description of physical and human resources may be appropriate, including staff CVs.

Costs

These may be estimates or precise figures, depending on the nature of the proposal.

The quality of your proposal will depend to a large extent on how good you are at seeing things from the point of view of the reader, your potential client.

Your reader will certainly want the proposal to be clear and readable. The point of your proposal, and its justification, must be made early in the document. A long proposal will need a summary at the front.

11.2 Specifications

A specification is a document that specifies. Engineers need to specify many things, and the use of the word varies from one branch of engineering to another. In software engineering, different stages in the development of a software product are based on particular specifications; these may include a requirements specification, a design specification and a test specification. They may be internal documents, or may be passed between departments or organisations.

In other branches of engineering, a specification is generally associated with an external contract, and it accompanies engineering drawings and other contract documents. It might specify the quality of work or materials, or specify the performance standards to be achieved. When the specification is part of a contract, it may have legal significance if there is subsequent disagreement over whether the contractor has satisfied its terms.

When you are specifying standards, you must remember that higher standards are harder to meet and therefore more expensive. It is a bad thing to specify quality standards that are unnecessarily demanding.

In all cases, the requirements or standards that are specified should be measurable and achievable. It must be possible to prove that they have (or have not) been met.

When writing a specification, you must use words carefully. Terms should be defined when there is any risk of ambiguity. However if there is no risk of ambiguity, then explanations of common terms are not needed – a specification is not usually written for a wide audience.

Specifications are cold factual documents but they must be readable. If they are hard to read, they may be misunderstood. Above all they must be unambiguous and complete.

11.3 Manuals

In industry, good manuals can be a selling point; they can increase customer satisfaction, improve standards of safety, and reduce the amount of after sales attention required. The manual goes with the product and to an extent determines its usefulness. The product could be a whole installation, a small device or a piece of software. Parts of the manual will be descriptions of the product, and parts will be instructions for using it. The proportions will vary with the nature of the product. The writer must distinguish clearly between description and instruction, and use appropriate language for each.

Considering the needs of the reader is important in all types of writing, but is probably more important for manuals than for any other type. You need to know a bit about the likely readers, including their level of technical knowledge, and their level of familiarity with the type of product. Students should clarify these details with their lecturer. At all stages of preparation, you must attempt to see the manual from the reader's point of view, and design the content to suit. It may be necessary for more than one manual to be produced in order to satisfy the needs of different readers. For example a software product might have a Training Manual for new users, a User

Manual for users with some experience of similar products, and a Reference Manual for those already familiar with the system.

You must aim to make your manual easy to use. This will be achieved through selection of material, good writing, informative diagrams, and helpful layout. The layout on the page should not be cramped, and should be designed so that diagrams are positioned conveniently, lists are clearly set out, and emphasis is given to the most important sections. A glossary may help if it is felt that readers may not be familiar with the definition of some terms.

You cannot write a manual unless you have detailed knowledge of the product. As a student, you are most likely to be asked to write a manual for a product you have developed yourself. You are not likely to be short of detailed knowledge; the extent of your knowledge may even cause problems since it may make it hard for you to see things from the point of view of the reader, who is likely to be a user without the same detailed knowledge. This makes writing the manual a demanding test of your ability to think and write clearly. One technique that could help you overcome this type of problem is to start preparing the manual while you are developing the product. You are more likely to see things from the user's point of view when you are facing the problems for the first time yourself.

Technical descriptions must be written in simple straightforward language, with reference to diagrams wherever appropriate. They should be kept separate from instructions.

Instructions tell the user what to do. The most user-friendly instructions are those that are the clearest to read, not the most informal or entertaining. Instructions must be thorough, and it is better to state the obvious than leave out something which is not obvious, especially when dealing with safety precautions. However including too many genuinely unnecessary things will risk alienating or boring the reader.

You must think carefully about language. Let us consider an example. At a particular stage in setting up a piece of monitoring equipment, a digital display must be set to zero using keys marked UP and DOWN. This adjustment is always necessary before taking actual readings. But how should we write this part of the instructions? Here are some alternatives.

The UP and DOWN keys allow the equipment to be zeroed.

This is a description not an instruction. Also **allow the equipment to be zeroed** is not precise; how will we know when it is zeroed?

It is necessary to ensure that the digital display is set to zero by pressing the UP and DOWN keys.

This is still not an instruction.

Press the UP or DOWN keys until the digital display reads zero.

This is an instruction, but while trying to be precise the writer has created a problem. This instruction can be followed literally without achieving the desired end. What if the digital display shows a number greater than zero and the user presses the UP key   ?

If the digital display shows a number less than zero, press the UP key until the display shows zero. If the digital display shows a number greater than zero, press the DOWN key until the display shows zero.

This is overdoing it! Surely we can assume that the reader can imagine the effect of pressing the UP or DOWN keys.

Set the digital display to zero using the UP or DOWN key.

This is better.

If you are required to write a manual as part of a student project, don't think of it as a writing chore, an anticlimax after the main technical challenges have been overcome. This is not writing for writing's sake, it is an integral part of achieving the goal of all your efforts – making something work in practice. Students sometimes underestimate the challenge of writing a good manual for their product.

In industry the challenge is not underestimated by the most successful companies. The writing of manuals is considered to be a highly skilled task, and, especially in software, one which is strongly linked to the eventual success of the product.

An effective test, and an instructive experience, is to watch as someone who is unfamiliar with your product attempts to use it for the first time by following your manual.

Further reading

For specialised advice you must refer to specialist books, and these will be recommended by the lecturers on your course. The books below concentrate on good writing rather than technical content.

Haslam, Jeremy M. *Writing engineering specifications.* E. & F.N. Spon, 1988.

Thirlway, Martyn *Writing software manuals – a practical guide.* The B.C.S. Practitioner Series, Prentice Hall, 1994.

All three books recommended in Chapter 10 contain advice on writing proposals.

12. Final year project reports

12.1 Final year projects and reports

In many engineering courses the final year project is the greatest challenge. A good project brings together the main themes of the whole course, including engineering science, engineering applications, specialised knowledge and design. It calls for talents that a graduate will need when entering the profession: judgement, technical understanding, originality and the ability to communicate well. Whether you do well or badly in the other parts of the course, a good final year project is a valuable achievement in its own right. It can, for example, help you find a good job, since your supervisor can praise it in a reference to a potential employer, and you can take the report (if finished) to the interview.

A successful project is a personal achievement, and that is the most satisfying aspect. You will go more deeply into this topic than any other. By the end, you will know as much about it as almost anyone else. You will make great demands of your own abilities. When you have finished, *you* can take credit for the achievements.

It is not possible to generalise about the content of the project; it could be virtually anything including research, development or design. You may have done work in a laboratory, at a computer, in a workshop, at a drawing board, out in the field, out in industry, or in the library. There may be some product: a prototype, a model, a piece of software, or a set of drawings; but the main output, in which all your thoughts, experiences, frustrations and commitment are "contained", is a report. No one will know what you did, how hard you worked, how much you achieved, if it is not described in your report. Your supervisor may be able to say on your behalf "I must say this student worked very hard . . .", but it won't count for much, believe me, when the other lecturers assessing the project reply "well, it isn't in the report". Final year projects are the ultimate proof of the importance to engineering students of the ability to communicate well.

You will have written quite a few reports in your course before you carry out your final year project. Yet writing the project report will be an especially challenging task. It is likely to be the longest report you will have written and potentially worth the greatest number of marks. It will need to have the qualities of a good engineering report as described in Chapter 10. It will need to be rigorous in an academic sense – with a thorough treatment of theory, an awareness of the work of others and a clear idea of what is genuinely original. The report may in fact be called a dissertation or even a thesis, but we will stick with the word report here. The advice in this chapter is to help you make it a good one.

12.2 Planning

Planning is one of the most important aspects of carrying out a project and writing the report. You must always have a clear plan which sets realistic targets and takes account of all the other demands on your time. You should start planning the report from the start of the project. Everything that you do, right from the first piece of introductory reading, needs to be represented in some way in the final report.

One of the worst mistakes is not to leave enough time for writing. This mistake can put you in a state of panic throughout the production of the report – not the best state of mind for one of the most important stages of the project.

In most final year projects, a student is supervised by a particular member of staff. It is important to make full use of your supervisor. You should regularly discuss your plan with your supervisor and start discussing the report well before it is time to start writing. You should seek comments on the draft of the first section that you write.

12.3 Contents

The general layout of your report should follow the pattern proposed in Chapter 10 (Reports). Refer back to the following sections if they are not fresh in your mind.

10.2 Structure
10.3 Beginning and end
10.4 Summaries.

Obviously the precise nature of the contents varies enormously with the subject of the project. It is not possible to predict the headings and subheadings that would be appropriate. However the report must cover the following aspects.

Relevance of the project

You will need to give some background information about engineering practice in your specialised area, and show how your project is relevant.

Aims

It is important to express the aims of your project briefly, clearly and precisely.

Theory

Most projects have some theoretical basis. Any fundamental statements of theory should be presented together in one section, not introduced when needed during the description of the project.

Previous work

Alternative subheadings would be **Literature survey** or **Review of relevant studies**. Someone, somewhere must have carried out work that is relevant to yours. You must describe their work and show how yours fits in. All relevant studies should be described together in this section. Later in your report, perhaps at the end of your analysis, you may wish to point out that your interpretation is similar to someone else's. But your reader should already be familiar with that person's work from your description of **Previous work**.

What you did

This should include any rejected options or ideas that turned out to be fruitless. Projects never go smoothly – you should describe the things that went wrong as well as the things that went right. This is partly to show your assessors how much work you did, and partly to help future investigators avoid making the same mistakes.

If your project is experimental, you will include the sort of information described in Chapter 8 (Laboratory Reports). You should give particular emphasis to interesting or unusual procedures and to safety measures. There should be enough detail to allow someone else to repeat your work.

You don't always have to describe your work in the order in which it was carried out. For example you might have tested something at one end of a physical range, then at the other end of the range, then in the middle. But the logical order for listing the tests, or presenting the results, would be ascending order of the physical parameter.

Analysis

This could be analysis of results or a critical review of what has been achieved. This may be the hardest part to write and also the most important. This is when the real thinking is needed and where the originality lies.

Conclusions and Recommendations

The conclusions must be related to the aims. You should make sensible, practical recommendations. These may include areas for further study. This can create a satisfying sense of continuity: previous work leading to your work, your work leading to future work.

Appendix

This is a good place to put lengthy sets of data, listings of software etc.

12.4 Writing

First of all, unless you have read it recently, look through 10.5.

Now let us consider the particular characteristics of writing final year project reports.

Level

You should write at a level to suit someone with your general knowledge of engineering but with no specialist knowledge in the area of your project. There is no need to explain basic engineering principles, but you must assume that your reader has no prior knowledge of your project. You are *not* writing for your supervisor.

General Style

Your style should be formal and precise. Every statement you make should be justified, and nothing you write should be vague. Use tables and diagrams wherever appropriate and no more words than are needed to make your meaning clear. Give references for all material that is not your own.

Describing what you did

A writing style for describing procedure has been recommended in Chapter 8 (Lab Reports). The emphasis should be on what was done, not who did it.

12.5 Technical information

Presentation of technical information is likely to be an important part of an engineering project report. Refer back to Chapter 6 (Technical Information) when you need to.

An aspect of final year project reports that is often disappointing is the integration of the technical data with the descriptive text. If a diagram is not referred to from the text, many readers will not see it. Most diagrams require some explanation; if you have not made the purpose of your diagram clear, there is no point in asking your readers to look at it. Remember, communication of technical information is a *partnership* between tables, graphs, diagrams and words.

12.6 References

In any report, and particularly an academic one, references should be handled carefully.

All the information in your report must come from somewhere. It could be data from your own experiments or computations, your interpretation of those results, someone else's results or interpretation, or an established principle. You may even be quoting someone else's words. Whatever the information, it must be clear to the reader where it has all come from. There are four good reasons for this.

1. *To support a statement.* To make it clear that the statement, if not based on your own evidence (as described in the report), is based on someone else's.

2. *To show how your work relates to other people's.* By demonstrating your knowledge of other work you show that you have made the most of what is available and taken care to avoid duplication.

3. *To allow your readers to find out more information by reading the publication to which you refer.* This means that you must tell them precisely how to find the publication.

4. *To acknowledge your sources.* To show that you are not pretending that someone else's ideas or words are your own.

Whenever someone else's work is involved you must make a reference. At the appropriate point in your text you give a reference code. The code allows your reader to refer to your list of references (near the back of the report). In the list you give the precise details of the publication to which you are referring.

The reference code is either a simple number written in brackets or as a superscript, or the author's surname and the year of publication.

The list of references has a different appearance for the two systems. In the number system they appear in numerical order (usually the order in which they are referred to in the text). In the name/year system they appear in alphabetical order of author's surname.

Example (number)

Good spelling may benefit engineers in their personal lives. Two-thirds of British engineers feel that if they improved their spelling they would raise their standard of living[1]. A survey by Foster[2] has shown that 80% of engineers who cannot spell "espousal" are divorced by the age of 35.

References (number)

1. James H.H. *Aspirations of the engineer.* Longman, 1995.
2. Foster G.E. The spelling of life. *Journal of Engineering Communication,* 1996, 8, 207–215.

Example (name/year)

Good spelling may benefit engineers in their personal lives. Two-thirds of British engineers feel that if they improved their spelling they would raise their standard of living (**James 1995**). A survey by **Foster (1996)** has shown that 80% of engineers who cannot spell "espousal" are divorced by the age of 35.

References (name/year)

Foster G.E. (1996) The spelling of life. *Journal of Engineering Communication,* 8, 207–215.
James H.H. (1995) *Aspirations of the engineer.* Longman.

If, in the name/year system (also called the Harvard system), you refer to two publications by the same author in the same year you distinguish them by lower case letters.

(James 1993a)
(James 1993b)

If there are authors with the same surname, the initials are included. If there are two authors for a particular reference, both surnames are given. If there are three or more, only the first name is given, followed by **et al.** (short for **et alii**, the Latin for **and others**).

(Smith D.C. 1992)
(Smith D.G. 1991)
(James and Alfson 1990)
(James et al. 1994)

The choice of system is up to you. The number system creates less disturbance to the flow of the text, and is probably more suitable for engineering reports and short papers. But if, while checking your work, you remember another reference which must be added near the start of your report, you will have to renumber all the later references if you want to retain the numbering sequence. For a PhD thesis, where hundreds of references are being collected and reshuffled over a period of years, the name/year system is more convenient. A final year project report falls somewhere in the middle. If you have more than about ten references I recommend the name/year system. Whichever system you choose, stick to it rigidly. If, for example, you use the name/year system, don't number the references as well.

The reference in the list must be precise and thorough. This is so that an interested reader can track down, from among all published material in the world, the particular publication that is referred to. Academics are rightly fussy about references, and tend to scrutinise them closely when they are assessing reports.

You must give the following information.

Journal article

author(s)
title of article
name of journal (italic or underlined)
year of publication
volume number (bold)
issue number (in brackets), if needed
page numbers

Book

author(s)
title of book (italic or underlined)
edition (if appropriate)
publisher
year of publication

Contribution in book

author(s) of contribution
title of contribution, followed by "In:"
editor(s) of book
title of book (italic or underlined)
edition (if appropriate)
publisher
year of publication
page numbers

Paper in conference proceedings

author(s) of paper
title of paper, followed by "In:"
title of conference proceedings (italic or underlined)
volume number (bold) or volume title
location of conference
year
page numbers

Report

author(s)
title (italic or underlined)
serial number
institution (name of institution, location)
year of publication

Thesis

author
title (italic or underlined)
degree for which submitted
institution, town and country if needed
year

Here are some examples. Note that the position of the year depends on the system used.

Number system *(in order: journal article, book, contribution in book, paper in conference proceedings, report, thesis)*

1. Fickson D.F. and Long S.E. Effect of infiltration on quality modelling in sewer systems. *Journal of Environmental Engineering*, 1999, 56(5), 421–429.

2. Chan C.L. *Developments in manufacturing systems*, 3rd edn. Macmillan, 1997.

3. Nichols K.H. Computer monitoring of unsteady pipe-flow. In: Khan D. ed. *Computer control and monitoring of engineering processes*. Wiley, 1998, 273–291.

4. Onof P. and Adenle J.B. Practical teaching of electronics systems design. In: *Proceedings of the 5th International Conference on Engineering Education*, 2, Naples, 1998, 397–402.

5. Jones D. and Evans D. *Survey of small engineering enterprises in Wales*, Report R72. Welsh Council for Engineering, Cardiff, 1997.

6. Walters P.R. *Recycling of construction materials*. PhD thesis, University of Devon, 1999.

Name/year system *(the same examples, but in a different order because they must be listed alphabetically)*

Chan C.L. (1997) *Developments in manufacturing systems*, 3rd edn. Macmillan.

Fickson D.F. and Long S.E. (1999) Effect of infiltration on quality modelling in sewer systems. *Journal of Environmental Engineering*, 56(5), 421–429.

Jones D. and Evans D. (1997) *Survey of small engineering enterprises in Wales*, Report R72. Welsh Council for Engineering, Cardiff.

Nichols K.H. (1998) Computer monitoring of unsteady pipe-flow. In: Khan D. ed. *Computer control and monitoring of engineering processes*. Wiley, 273–291.

Onof P. and Adenle J.B. (1998) Practical teaching of electronics systems design. In: *Proceedings of the 5th International Conference on Engineering Education*, 2, Naples, 397–402.

Walters P.R. (1999) *Recycling of construction materials*. PhD thesis, University of Devon.

Quoting

If you wish to use someone else's precise words you must enclose them in inverted commas and give the reference. Here are two examples.

> Fickson and Long (1999) point out that "in some catchments infiltration can have a significant effect on sewer-flow quality".

> "It is important that measuring instruments do not significantly disrupt the flow in pipes." (Nichols 1998)

Quoting in this way can be effective, but there are limits. If substantial chunks of your report are not written by you, there will be less of your own work to earn marks. Quoting without inverted commas and a reference is **plagiarism** (using other people's words or ideas as if they were yours). In academic terms that is stealing, and constitutes a crime.

Bibliography

A list of references is precisely that – a list of publications to which you have referred in your report. A bibliography is a list of publications that you have used during your study, or that you recommend. You might write comments on the items in a bibliography. The **Further reading** recommendations in this book make up a bibliography.

Checklist for a final year project report

- for each table, graph and diagram, use the checklists in Chapter 6
- if the project includes laboratory work, use the checklist on p 82

- use the checklist for format of a major report on p 96
- use the checklist for text of a report on p 98
- use the checklist for content of a report on p 104

- have you written a good summary?
- have you shown the relevance of your topic to engineering practice?
- have you included photographs where appropriate?
- have you related your study to the published work of other people?
- have you quoted or referred to other people's work in a clear and consistent way?
- have you used the recommended format for references?
- has your supervisor checked your report, and have you acted on the suggestions?

12.7 General points

The assessment of many final year projects also includes a spoken presentation or an interview. Advice on spoken presentations is given in Chapter 13. Interviews for projects are included in Chapter 17 (Interviews).

13. Spoken presentations

Most engineering students give spoken presentations during their courses. Likely occasions are seminars, and presentations on design work and projects. The audience usually consists of fellow students and members of staff.

This chapter is designed to help you give spoken presentations as a student, when you probably have little experience of this sometimes daunting activity. It covers your state of mind, visual aids, and preparing and giving the presentation.

13.1 State of mind

Good English is important when giving a presentation just as it is when writing; but the main problem with speaking is not planning what to say, but managing to say what you have planned once the audience is in front of you.

Most engineering students realise that spoken presentation skills are worth developing, but very few look forward to the opportunity. The main reason is a general feeling of apprehension which tends to be called "nerves".

There is nothing wrong with being nervous. Professional performers do not try to eliminate the feeling; it carries out the important function of focusing energy and determination. However, nerves can be a nuisance if they affect your memory or voice or manual dexterity, and that (the nuisance not the nerves) needs to be overcome.

There are techniques for controlling nerves: relaxation exercises, conscious breathing, smiling at yourself in a mirror. We all need a way of settling ourselves, even if we've been giving spoken presentations for years. I cannot start a lecture without holding a pen in my right hand (even if I don't need to write anything).

But there is only one truly effective way of overcoming the effects of nerves – **preparation**. Every good presentation is a well-prepared presentation.

13.2 Visual aids

Think of the television news. Even the most straightforward comparison is illustrated with some kind of visual image. The value of the pound has risen against the dollar by 0.1 of a cent. That is an easy statement to understand, and yet a simple illustration helps. We understand better when we see things as well as hear them. Visual aids can also help you in your presentation by:

> emphasising your points
> creating variety
> taking the pressure off you, by causing the audience to look at the screen instead.

Here are the types of visual aid that are likely to be available to you.

Overhead projector (OHP)

This is probably the device you will use the most. An overhead projector is a simple machine. One ON/OFF switch controls the light and a cooling fan. Focus is adjusted by moving the reflecting head up or down the supporting column. The image can be moved vertically on the screen by adjusting the angle of the reflector. To change the size of the image or its horizontal position, the whole projector must be moved. An overhead projector can be used in normal lighting.

The transparencies are usually made from clear A4 sheets. They can be produced in a number of ways: by hand using special pens, by photocopying (or thermal copying) from a paper original, or by computer. A computer can also be connected directly to a "tablet" positioned on the OHP or to a projection system, removing the need for transparencies. (In this case the room usually needs to be darkened.) Whichever method you use, you should make the material large enough to be seen clearly, and not include too much information on each image.

If you are making transparencies by hand you must use OHP pens. Don't use a thin point as the writing will be hard to read. You should use washable (rather than permanent) pens if you want to reuse your slides or correct mistakes easily.

Material can be photocopied on to clear sheets ("acetates"); this could be a diagram that you already have on paper (perhaps enlarged on the photocopier). You can create material using large lettering from a word processor or from a computer package for producing visual material. If you photocopy ordinary-sized typing it will not be large enough to be read clearly. The acetate sheets that go into photocopiers have paper backing-sheets attached. These are removed before the transparencies are shown, and you should do this before your presentation. Tearing the backing-sheet from each acetate during a presentation can become irritating to the audience.

Overhead transparencies based on computer lettering usually look far more professional than those that are hand written. Presentations illustrated directly from a computer can be even more professional.

Slide projector

This is for showing 35 mm or "2 × 2" slides. (35 mm is the overall width of the film, and the square mounting has 2-inch sides.) You are most likely to use 35 mm slides for photographs. Keywords and diagrams can also be placed on 35 mm slides, but in a student presentation an OHP is more likely to be used for this sort of material. A slide projector needs the room to be darkened.

It is likely that you will use a projector in which the slides are placed in a circular "carousel". A surprising amount can go wrong with these machines and it is worth spending some time making sure you understand how they should be operated. The slides are placed upside-down, but the right way round, in numbered slots in the carousel. Before you start to load, you should make sure that the slit at the bottom is lined up with the slot marked "0". Never force the carousel into or out of position (however anxious you are).

The slides are usually shown using a remote control which allows you to move forwards or backwards through the slides, and adjust the focus. Learn which button to press; constant mistakes with slides can be annoying for the audience and distressing for the presenter.

Video

You are likely to show only short sections of video in a presentation. If it is a video that you have filmed yourself (of a demonstration or experiment) it will probably be shown with no sound while you talk over it.

Boards

There will probably be a vertical surface for you to write on, although using it may not be a good idea as we will discuss shortly. You are most likely to use the board when you are answering questions. It may be a **white board**, in which case you must ensure that you use the correct pens, or a **chalk board**. An alternative is a **flip chart**, consisting of large paper sheets which can be flipped over when full. A flip chart is not likely to be used in a room which contains a board unless there is some need to refer back through the sheets.

13.3 Preparing

Notes

If you become a cabinet minister or a celebrity you may read speeches word for word. This may be because it is crucial that you do not risk saying a single wrong word, or because someone else has written the speech for you. You will by then have mastered the difficult art of reading out a whole piece and still making it sound interesting and sincere. Alternatively if you have to give a speech at a wedding, or a speech of thanks for an important award, you may memorise every word of what you plan to say.

For a student presentation neither reading nor memorising is likely to be appropriate. Reading is not appropriate because your audience will quickly become bored unless you are an expert at reading. It is very important that you look at the audience while you are speaking, and if you are reading from a script you cannot be looking up at the same time. Memorising every word is not appropriate because your delivery will sound unnatural, and you will run the risk of disaster in the event of memory-loss.

So your presentation must be based on notes of what you plan to say. Your notes must be detailed enough to prevent you from leaving out something important, but brief enough to allow you to look at the audience while you are speaking. Your notes must be prepared with great care – they will be your main support during the presentation.

You may also want to illustrate your presentation with transparencies for the overhead projector giving subheadings and keywords. Some people combine this overhead projector material with their personal notes (after all, both are mostly made up of subheadings and keywords). They do not use paper for their personal notes, they prepare overhead projector transparencies with the dual function of reminding them of what they want to say and acting as a visual aid for the audience. This is a method well worth considering for student presentations. For the method to be successful, the material for the OHP must be sufficiently detailed, and your practice sufficiently thorough, for you to be reminded of everything you want to say (even when you are nervous). Figure 13.1 gives a sample slide of subheadings and keywords, together with what might actually be said during that part of the presentation.

(Spoken:)

One of the main problems with the water depth gauges is calibration. This is also a problem with the traditional twin-wire wave probes. The gauges themselves measure the conductance of the circuit formed with the water, and the calibration of a gauge is really the relationship between conductance and water depth. This is found by taking readings at a series of steady water depths.

The conductance of the water itself is affected by temperature and chemical composition. If these vary during a series of tests (as they usually do), the calibration of the gauges will become inaccurate. The gauges therefore need frequent recalibration, and this is time consuming.

This was a problem with the gauges in my experimental set-up. The time period between recalibrations varied during the day. In the morning of a typical day, three or four recalibrations were needed – even if the equipment had been run for the first hour without taking any readings. In the afternoon, recalibration was not normally necessary.

I tried to investigate the reasons for the morning variation. During one morning I measured the temperature of the incoming water and found that it did not change significantly, even though the gauges needed to be recalibrated as normal. I have not been able to work out the reasons for this. I think the most likely explanation is that the chemical composition of the water in the sump varies because deposits are being disturbed.

Problems – CALIBRATION

Relationship between conductance and depth

Temperature, chemical composition
– frequent recalibration

Variation during day – reasons?

Figure 13.1 Keywords on an OHP transparency, and what might be said

Preparing content

Plan your content using the "writing process" approach described in Chapter 5, with the three main stages:

ideas – sorting – structuring.

Think carefully about the beginning and end. It is often a good idea to prepare an introductory overhead transparency giving your name and the title of your presentation. Even if someone introduces your presentation and gives your name and subject, it is still worth showing this first overhead. You will say something like "here again is the title of my presentation". There is nothing wrong with the repetition; the audience may not have taken it in the first time.

Think about what your audience needs to know. Remember: you know something that they don't – that's why you are giving the presentation. In your introduction, give an outline of the structure of what you are going to say.

Think of a positive way of ending. At the very least say "That concludes my presentation. Thank you." The worst way of ending is a pause during which you realise there is nothing more to say, followed by ". . . well . . . that's it".

Planning visual aids

Plan what you will say about every piece of visual material in your presentation. When showing photographs as slides, be selective about which slides to show. Don't include a slide that does not show anything new, or which is of poor quality. Prepare all your visual material in advance. If you try to write on the board during your presentation, you will cause unnecessary delay and your writing may be hurried, nervous and hard to read.

Practising

Try to practise as much as you can. You can practise at home, speaking quietly and looking at your overhead transparencies on a white sheet of paper on a table. Think about how long each overhead should be shown. Time the whole presentation and try to pace yourself.

Try also to practise in the room in which the presentation will be given (or a similar one). Find a position at the front of the room where you are comfortable. Stand on the side of the screen that allows you to point with your best arm, and work out where on the table to leave your overheads or notes. Do this with the overhead projector switched on to make sure that your papers are not blown away by the current of air produced by the fan. Practise using any other visual aids.

When the big day arrives, get to the room well before the session starts and have one more think about where to put things and where to stand. If you are using a slide projector, put your slides in the carousel (or make sure they are ready to go in if someone else is using it before you). If you can, check the focus and position on the screen of the first slide. If you are using a video monitor, make sure that it is in a convenient place, with the volume set to the appropriate level and the tape at the right position.

Checklist for preparing for a spoken presentation

- have you thought about what the audience needs to know?
- have you prepared notes (on paper or transparencies) which will guide you through your content but not prevent you from looking at the audience?
- have you prepared appropriate visual aids to enhance your presentation?
- have you worked out what to say about each transparency/slide?
- have you tried out your visual aids?
- have you practised and timed your whole presentation?

... on the day ...

- have you checked the room and the equipment?
- if you have 35 mm slides, are they ready to go into the projector?
- are your notes and transparencies in order?
- do you have a watch?

13.4 Giving the presentation

Basics

Unless the session is very intimate, stand up. Maintain good eye contact with the audience. Don't spend the whole time looking down at your notes, or backwards at the screen, or down to the top of the overhead projector. It doesn't matter who you look at; look at someone, anyone, and keep looking around the audience; try in effect to look at everyone. Of course you will have to look away to operate visual aids or read notes, but then look back.

Don't expect too much from the audience; they won't all be nodding enthusiastically at all your points. This is one of the things you will notice most when you give your presentation – there is not the feedback of normal conversation. Yet you should still look for subtle signs from the audience that your message is or is not getting through.

You must aim to be heard clearly at the back of the room, but don't shout. Try not to speak in a monotone, but don't give your voice unnatural intonation. Most people speak too quickly when they are nervous. If you try consciously to speak slowly it will give you time to think, and allow you to be heard more clearly. If you need to pause, do so. You can pause for longer than you think without worrying the audience.

Show enthusiasm, try to be yourself, smile when it is natural to do so. A little humour may even be appropriate, but bear two things in mind. You should never *expect* to get a laugh, otherwise you will be put off if you don't. But if you do get one (hopefully when you wanted one) follow it with something serious.

Don't try to be artificially formal or informal. The formality or informality of the occasion will look after itself.

Using visual aids

Don't put up a beautiful diagram or table (that has taken you hours to prepare) for just a second and then remove it. You should leave enough time for the audience to understand the content. To simply leave a silent pause would probably make you feel uncomfortable, so the best idea is to describe or read what is being shown. The audience are more likely to absorb the information if they see it and hear it at the same time. Think of sports results on the television. They are shown on the screen and read out in full, yet you don't say "oh shut up, I can read it for myself".

When you are talking about an overhead you must be careful not to stand in the way: neither blocking the light shining on the screen, nor preventing one section of the audience from seeing the screen. After you have changed a transparency, it is best to look quickly at the screen yourself to check the focus and position.

If you are showing a diagram, you may wish to point at certain features. One way is to point with a pen, or special pointer, on the surface of the transparency itself. The profile of the pen or pointer will then show up on the screen. If you are planning to use this method, consider three things.

1. If your hand is shaking slightly this will be magnified on the screen.
2. There is a danger that you will be obstructing the audience's view.
3. You may find that looking down into the light dazzles your eyes.

The alternative, which I prefer myself, is to point with an arm at the screen.

This allows you to make more open gestures, and communicate enthusiasm for what you are showing: "we had to design a special device to go **here**" or "agreement between the readings is good **here**". It also means that you and the audience are looking at the same thing: it is a shared point of attention. This technique is only suitable in small to medium sized rooms. In a large room in which the screen is so large that you need a stick to point with, the possibility of the shaking hand will need to be considered again.

Another way of focusing attention on an overhead is to reveal subheadings progressively by drawing a piece of paper or card down the transparency. This is quite a hard thing to accomplish. The three disadvantages of pointing at the surface of the transparency apply, and there is also the problem that when you have got more than halfway down, the piece of paper will no longer balance on the surface of the overhead projector and unless held firmly in position will slip down or be blown off by the air from the fan.

If there is a period in which you do not need to use the overhead projector, switch it off.

Don't attempt ambitious combinations of visual aids. It can sometimes be effective to show overheads and 35 mm slides side by side, but in many rooms it isn't possible. When you need to swap between one and the other on the same screen, you should try not to do so too often. The OHP is much brighter than the 35 mm projector, so if you leave spaces in the carousel between groups of slides, you can show overheads without switching off the 35 mm projector. Better still, place blank (opaque) slides in the spaces. You must however switch off the OHP before showing slides on the same part of the screen.

Timing and pacing

This is one of the hardest aspects of all. First of all, it is easy to forget to look at your watch when you start. Also it is easy to say more than you planned, and then realise right at the end that you still have a lot to cover. In that situation the best thing is simply to drop some of your material. Do not try to speak very quickly, or keep saying "I'm sorry, I know I'm running out of time, but I really must describe . . .".

There are no special techniques for time-keeping. Just prepare carefully, practise, and keep an eye on the time.

Style

Dress depends on the occasion. You may feel a little uncomfortable if you are dressed very smartly and nobody else is, but you will feel worse if you are the only person who is not dressed smartly. Feeling smarter than usual will help to put you in the right state of mind.

Don't stand rigidly. Move around and gesture in a natural way. Keep your approach simple and direct. Don't get too concerned about controlling body language or audience psychology. It may be time to work harder at gestures and refinements to technique when you have more experience. As a student, if you have something interesting to say, and you have prepared well, your presentation should be a success.

Group presentations

All of the material in this chapter is relevant to group presentations. In addition you must make sure that a group presentation is well coordinated. The introduction should name the members and define what each will cover. There should be a prepared handover between group members: "I will now hand over to Ian who will give more detail on the planning of the tests". Use each other's normal familiar names, do not try to be artificially formal by referring to each other as Mr . . ., Miss . . . etc.

The concluding part should draw together the themes of the group presentation as a whole.

Being videoed

You can learn a lot from seeing a video recording of your presentation.

During the presentation you should try not to be distracted by the fact that you are being videoed. Concentrate on communicating with the audience, not on the camera.

At playback time remember that most people hate hearing their own voice if they are not used to it. Self-criticism is easy; to identify all your good points you will probably need someone else, preferably your lecturer.

Answering questions

Responding to a question that you can answer well will be the most relaxed part of your presentation. This is your reward for "knowing your stuff".

If you are asked a question which exposes an area of weakness, don't try to bluff. The best thing is usually to make comments in a related area where you do have some knowledge, but to make it clear that that is what you are doing.

Checklist for self-assessment of a presentation

- did you look at the audience?
- did you feel that your presentation interested them?
- did you communicate your enthusiasm for the subject?
- did you make good use of your visual material?
- did you finish on time without rushing?
- did you say most of what you wanted to say?
- did you answer questions well?
- what would you do differently next time?

14. Communication in context

The previous chapters have considered basic elements of the use of English, followed by a number of formal ways of communicating. Now is a good time to reflect briefly on how formal communication fits into the working life of engineering students and engineers. Here are three general principles. They are listed first, then considered one by one.

1. *Communication is just one part of being a good engineer or engineering student.*
2. *A great deal of communication is not of the formal type emphasised in this book.*
3. *Communication is an important element in working with people.*

1. Communication is just one part of being a good engineer or engineering student.

Good communication will help you to become a good engineer, but there is a lot more to being a good engineer than simply being a good communicator. This book is certainly not intended to encourage you to develop an artificial style of communication in which **how** you say something is more important than **what** you say. The aim is to help you develop a clear style so that when you have something to say you can say it effectively. Fortunately, engineering is a profession in which genuine achievement is more important than the appearance of success. You cannot disguise poor engineering by slick presentation – not for long anyway. Good communication in engineering is a means to an end, not an end in itself.

2. A great deal of communication is not of the formal type emphasised in this book.

In spite of their importance, reports and spoken presentations are not the only means of communication for engineers and engineering students. A great deal of information is communicated informally: in conversation, as scribbles on scraps of paper, or on the phone. We should not try to apply the same principles to informal communication as we do to formal communication. We would have a strange conversation if we avoided using all unnecessary words. If colleagues say "Good morning, how are you?", and you reply "Please use only necessary words", they will think you are pretending to be a robot.

Much communication does not involve words at all. Gestures, posture, facial expression and tone of voice can communicate a great deal. Some communication does not aim to get across a message but to establish a good working relationship or inspire confidence.

3. Communication is an important element in working with people.

Engineering students frequently work in groups, for example during laboratory classes, design exercises and projects. Group work can be frustrating and it can be stimulating. Groups can sometimes devise outstandingly imaginative solutions to complex problems, and sometimes waste hours on something very simple. One of the main factors in determining the quality of a group's performance is communication. A sign that a group is communicating well is that it is good at deciding when the members should work together as a group and when they should divide up the workload between them.

When working in a group, an important communication skill is listening; after all, during discussions in a group of four you may be listening three times longer than you are speaking. Listening well involves respect for other people's ideas, willingness to encourage them to make their contributions, and concentration. Listening, in a different context, is also an important study skill for students.

In industry, engineers often work in multi-disciplinary groups. This makes particular demands of their communication skills in every respect, and engineers are not always good at communicating with people outside their discipline (remember, good communicators are not just good at giving out messages but also good at receiving them). These groups are often led, to many engineers' disappointment, by non-engineers. I discussed this with a quantity surveyor who was leading a project team which included engineers. When I asked him what he thought he was better at than the engineers, he said "listening".

Working with other people cannot always be based purely on friendship and cooperation. It is often important in working relationships to know how to be **assertive**. You are assertive when you seek mutual respect. The respect you show others is as important as the respect you expect from them. You should not be aggressive or bossy, nor should you be passive or defensive.

The experience that you gain from working in groups while you are a student is a valuable preparation for a career as an engineer. Employers are attracted to graduates who can demonstrate success in group work.

Further reading

Stanton, Nicki *Communication*. Macmillan, 1990. This book contains a lot of interesting material on broader aspects of communication: the process of communication, listening, non-verbal communication and working in groups.

Back, Ken and Kate Back *Assertiveness at work*, 2nd edn. McGraw-Hill, 1991. A comprehensive and interesting book.

15. Letters

15.1 Format

There are well-established conventions for the way a letter is set out. If you do not follow them you will risk losing your reader's confidence. General appearance is also important; an untidy letter may not be taken seriously.

Here is an exchange of letters. Angela Fulchrum is a final year student on the BEng Manufacturing Systems Engineering at the University of Devon. She is carrying out a project on the modelling of manufacturing efficiency. She feels that she should visit some small to medium sized manufacturing enterprises to gain some firsthand knowledge. She has written letters to five local companies, asking if she could visit their plants. Here is one of the letters, and the subsequent correspondence.

18 Bere Crescent
Souweston
Devon SU8 2ER

25 January 1996

Mr D. Greene
Manufacturing Manager
Medium Makers
Halo Industrial Estate
Burridge
Devon BU7 4RT

Dear Mr Greene

Final year student project

I am a final year student on the BEng in Manufacturing
Systems Engineering at the University of Devon. I am
carrying out a project on aspects of the modelling of
manufacturing processes. I am keen to visit local
manufacturing plants in order to increase my
understanding.

I would be grateful if you would allow me to visit your
plant. I appreciate that you would want such a visit to
be brief, and I would try to ensure that I did not take
up more than one hour of your time. The most suitable
time for me to make a visit would be during this term:
between now and 2 April.

If you are able to agree to this request, please could
you suggest a time and date.

I will not make any reference to your plant in my report
without your permission. I would be happy to give you a
copy, if you wished to have one, when the project is
complete.

Yours sincerely

Angela Fulchrum

MEDIUM MAKERS,
Halo Industrial Estate, Burridge, Devon BU7 4RT
phone: 012121 666444   fax: 012121 666445

Our ref: DFG/hj

6 February 1996

Angela Fulchrum
18 Bere Crescent
SOUWESTON
Devon SU8 2ER

Dear Angela Fulchrum

Visit to Burridge plant

Thank you for your letter of 25 January. You are welcome to visit the plant at 2.30pm on 20 February. Phone me if this is not convenient so that we can discuss alternative dates.   •

If you are coming at this time, please could you confirm in writing. Perhaps you could give me a few more details about your interests so that I can make sure you see the relevant parts of the plant. I also need written confirmation that you will be covered by your own institution's insurance arrangements.

Yours sincerely

DGreene

Derek Greene
Manufacturing Manager

18 Bere Crescent
Souweston
Devon SU8 2ER

11 February 1996

Mr D. Greene
Manufacturing Manager
Medium Makers
Halo Industrial Estate
Burridge
Devon BU7 4RT

Dear Mr Greene

Visit to Burridge plant

Thank you for your letter of 6 February (your reference DFG/hj). I am very grateful to you for agreeing to my visit, and I confirm that I will arrive at 2.30pm on 20 February.

My project is a study of a novel method of modelling manufacturing efficiency. I am applying it to predicting the impact of manufacturing cells. I would particularly like to see the layout of your plant and the movement of work around it. However my visit will be of great value to me irrespective of the specific layout that you use.

I will be covered by the University's insurance while I am on my visit.

Yours sincerely

Angela Fulchrum

Let us consider the format of letters in detail, with reference to the examples above.

Your address

It is normal to have no punctuation in an address. Crescent *could* be abbreviated to Cres, Road to Rd, but why? To save less than a second of your time? If you are typing your address, it usually goes in the top right of the page. If you are using headed notepaper, you will not need to type your address.

Date

If you have typed your address in the top right, the date usually goes underneath it. With headed paper, the date can be written on either side (it might depend on the design of the stationery). I recommend that you write the date with the format: 11 February 1996. There is no need to write 11th. The month and year written out in full looks better than 11 Feb 96. The American convention is to write February 11, 1996. That is a good reason for not writing 11.2.96, which could be understood by an American to be November 2.

Their name and address

You would not include the recipient's name and address in a letter to a friend (except on the envelope). But it is important to include it on a formal letter so that you have a record, on the copy that you keep, of where the letter was sent. Also formal letters are often sent in envelopes with windows, with the notepaper folded so that this name and address is visible when the envelope is sealed (and does not have to be printed again on the outside).

The normal position for the recipient's name and address on the notepaper is on the left, above the Dear . . .

Dear . . .

Angela had done her homework properly and found out the name and precise job title of the manager in charge of manufacturing. If she had not known the name she would have had to start her letter Dear Sir/Madam.

When Derek Greene replied, he knew that he was writing to someone called Angela Fulchrum. He could not write Dear Mrs Fulchrum or Dear

Miss Fulchrum as he did not know whether or not she was married, and didn't think it was relevant. Dear Angela might have seemed over-familiar or patronising. He could have used the safe, fairly formal Dear Ms Fulchrum. He preferred Dear Angela Fulchrum. This is a useful alternative, and is particularly helpful when someone has given only surname and initials or has written their first name as Chris or Nicky (and therefore could be male or female), or has a name whose form is unfamiliar to you.

If you are replying to someone whose name is printed below the signature on their letter as, say, E. Stuart (Mrs) or Dr J. Robinson, you should write Dear Mrs Stuart or Dear Dr Robinson.

Heading

A heading (like **Visit to Burridge plant**) helps to clarify the purpose of the letter. Notice that Angela in her second letter used the same heading as Derek Greene, so that he would recognise the topic more quickly. The heading is written below the Dear . . ., and should be bold or underlined.

Opening

If your letter refers to previous correspondence you should make this clear. If you want to be coldly neutral (for example about a final demand for payment), you can start with **I refer to your letter of 11 February.** Derek Greene may not have much spare time for showing students round his plant, but having decided to agree to Angela's request he begins his letter with a warm **Thank you for your letter.** The reference code is a convention of the particular office. Angela was right to quote the reference in the way that she did, and there was no need for her to make one up for her own letters.

After references to previous correspondence, you should come to the point of your letter quickly.

Closing

There is no need to "round off" a letter with a special closing remark.

Yours . . .

Letters that start **Dear Sir/Madam** end **Yours faithfully.** Letters that are addressed to someone by name end **Yours sincerely.**

15.2 Style and length

The language in letters should be clear and concise. There is no special "letter language", or "business language"; you should never <u>try</u> to be formal or you will become pompous. Nor should you be over polite or obsequious. Angela wrote that she was "very grateful" because she was; but if you say that you "would be very grateful" too many times in a letter, you will sound insincere. However, you should be careful not to use a style which is so informal that it sounds casual.

You should make your letters no longer than is necessary. Most letters fit on one page. If you have a great deal of information to communicate, it is usually better to present the information in a self-contained note or report, and write a covering letter which explains what the note or report is about and why it is being sent.

Checklist for preparing for a letter

- have you included: your address
                       the date
                       their name and address
                       a heading
                       their reference (if appropriate)?
- have you followed the convention:
                       Dear (name) – Yours sincerely
                       Dear Sir/Madam – Yours faithfully?
- does your letter come to the point quickly?
- is it concise?

---

16. CVs and job applications

CVs and job applications are important tests of communication skills; that is why they are considered in this book. This chapter is designed to help you write CVs and job applications, but it is not meant to be a comprehensive guide to finding a job. For the full story you should listen to the guidance that is given by the careers advisers at your institution.

They will explain that there are several essential stages before a job application becomes a writing task, including analysing the job description, finding out about the employer, and thinking about which of your attributes are likely to be of most use in the job.

In Chapter 1, I referred to engineering employers' low opinions of the communication abilities of graduates. In all fields employers tend to be critical of students' skills in making job applications. So if you work hard at your application, it may easily shine out above the rest. Pay careful attention to all aspects, including overall content, detail and quality of presentation. Your application – its appearance, and the care, thought and imagination that have gone into its preparation – can really help you get an interview. When there are many applicants with similar qualifications, the impression given by the application may be all an employer has to base decisions upon. You should always try to see your application from the employer's point of view.

You will want to make the most of what you have, and to present it in the best light. If you think there is something special about you, you must make sure that it is obvious in your application. But do not try to make something that is not remarkable sound as if it is, as in the following extract from a student's application.

Having studied Engineering for three years I believe the knowledge I have gained would be a very considerable asset to your organisation.

In referring in this way to his degree (a qualification which all applicants for a graduate engineer vacancy would have possessed) I think this applicant was trying too hard. Yet the same person had given no prominence to the highly relevant fact that he had run his own business for two years before going to university.

A CV is normally used for a job application when an application form is not mentioned in the job advertisement (the wording might be "applications should be made in writing"), or when your application is not in response to a specific advertisement. The contents of a CV and an application form are similar. Let's consider the CV first as it is a greater test of communication ability. Many of the points will also be relevant to application forms.

Whatever format your application takes, keep a copy of what you send for your own records, and to help you prepare for an interview.

CVs

CV is short for *Curriculum Vitae*, the Latin for "the course of life". The word **résumé** is generally used in North America.

We met Angela Fulchrum, final year engineering student at the University of Devon, in the last chapter. Her CV is shown on the next page.

Angela's CV, like most, consists of bare facts without continuous prose. Yet it has been carefully thought out in order to emphasise her suitability for a particular job. The job that she is applying for is in general engineering and so she shows that she has some experience in practical engineering through her vacation jobs. It would not be wise to try to give any more prominence to these, since the periods were short. If she were applying for a job with a significant administrative component, she might give more information about her duties as Secretary of the University Harriers. Each time she uses her CV she will think about the appropriate emphasis. We talk about "Angela's CV" as if it were a definitive, unchanging document, but that should not be the case. Giving the right emphasis is one of the subtle arts of CV writing, and you must listen to advice to suit your own circumstances.

When you have an idea of overall strategy, think about the layout, and be fussy about everything. In presentation terms your CV must be as perfect as you can make it. A spelling mistake would be disastrous (and remember, computer spell-checks do not pick up all mistakes).

If something is given a lot of space it tends to appear important to the reader. So Angela did not list her GCSEs in a vertical column, or give grades, as that would have given more prominence to them than to the later stages of her education.

Most students have some work experience by the time they are applying for permanent jobs. Vacation work in engineering, or better still the industrial training component of a sandwich degree, are obviously the most relevant to jobs in engineering. It is also worth including non-engineering work experience, especially if it involved some responsibility. If you include a mixture of types of work experience, think carefully about the emphasis you give to each.

Angela Jane FULCHRUM

ADDRESS	**Home**	**Term**
	(3-22 April, 1 July-)	(23 April-30 June)
	54 Pampard Road	18 Bere Crescent
	Croydon	Souweston
	Surrey CR2 4PR	Devon SU8 2ER

TELEPHONE 0181 432 54 77   01111 111456

DATE OF BIRTH 22 October 1974

NATIONALITY British

EDUCATION AND QUALIFICATIONS

1985 to 1993	Haling High School, Croydon
1991	GCSE (7): Maths, Physics, Chemistry, English, French, Geography, History
1993	A-levels: Maths(B), Physics(C), Geography(C)
1993 to 1996	University of Devon BEng(Hons), Manufacturing Systems Engineering Final year project on modelling of manufacturing efficiency

WORK EXPERIENCE

1994 (6 weeks)	Laboratory Technician (Electronic Eng) University of Manitoba, Canada Maintenance of equipment, assistance to research students undertaking practical work
1995 (7 weeks)	Design Technician Dufin Engineering plc, Croydon Design of small-scale alterations to hydraulic machinery for new applications

SKILLS   Computing: experience of WordUp, Project+, Super-CAD/CAM, C programming
Clean driving licence since May 1993

LEISURE INTERESTS

Sport : Athletics
Secretary, University Harriers
Music: Singer in folk group

REFEREES

Dr G H Bowen	Mr I Saines
BEng Course Leader	Senior Engineer
School of Engineering	Dufin Engineering plc
University of Devon	Centre House
Campus Park	Albert Street
Souweston SU1 1AA	Croydon CR1 0HB
01640 453967	0181 549 3028

Angela's CV is set out in normal chronological order. It is just as appropriate to set out a CV in reverse order. This tends to give more emphasis to recent achievements, which is often a good tactic. Be consistent: don't give your education in one order and work experience in the other.

Leisure interests can help to show that you are an enterprising or interesting person. But if they don't show that, don't include them.

You will need to name people who can write references for you. The normal number is two, but some employers may ask for three. One should be a lecturer on your course, preferably Personal Tutor or Course Leader. Ideally another should be your boss during a significant period of work experience.

Ask for permission first before you give someone's name. You want your referees to have a good opinion of you at the time they write the references, and an unexpected request from an unfamiliar source may threaten this state of mind. Obviously you should choose someone who you think will write you a decent reference. (So it *does* matter what your lecturers think of you!)

Make sure you give the details of your referees' names (with Dr, Professor etc.), job descriptions and addresses correctly. Give their phone numbers if you can.

You should prepare your CV on a word processor with a good quality printer. It may fit on to one page (like Angela's), but if it would be cramped on one page, give it two. Lay out the text so that it fills up both pages, and make the break between the pages in a suitable place. Submit the two pages stapled together, not a double-sided sheet. A long CV is not "better" than a short one, especially if the long CV is full of padding.

Get someone to check over your CV – a fellow student, careers adviser, or lecturer, or preferably all three. Spend time responding to their comments. If you feel reluctant to show it to other people, it must need some more work. Most importantly, look upon these brief notes on CV writing as the beginning, not the end, of your education in CV preparation.

Letter of application

You will need to send a covering letter with your CV. Angela's is shown on the next page.

She has highlighted some of the most relevant points in her CV, without being repetitive.

Application forms

Many of the comments already made about CVs apply to application forms. Opportunities for giving emphasis to particular aspects of your background still exist, but there is less scope.

18 Bere Crescent
Souweston
Devon SU8 2ER

11 May 1996

Mr G. Parland
Personnel Manager
JJ Engineering
Enterprise Road
MANCHESTER M16 7TY

Dear Mr Parland

Vacancy for Graduate Engineer

I would like to be considered for the post of Graduate
Engineer (reference E/563) as advertised in New Engineer
magazine on 6 May. I attach my CV.

I will complete my BEng (Hons) in Manufacturing Systems
Engineering at the University of Devon in July. In my
final year I have been carrying out a project on the
prediction of efficiency resulting from the introduction
of manufacturing cells. This has increased my interest in
the practical aspects of manufacturing engineering. I also
have practical engineering experience from two vacation
jobs.

Yours sincerely

Angela Fulchnum

There is a Standard Application Form (SAF) for graduate vacancies, which is accepted by some engineering employers. Copies, and advice on completion, will be available from your careers service. Other companies require applications to be made on their own forms. If you have been asked to apply on an application form, don't send a CV instead, or fill in cursory details on the form and attach a CV.

You may be required to complete the form in your own handwriting, in which case you should write as neatly as possible. If you are allowed to type the form, you will achieve better quality by doing so, provided you have access to a good typewriter and know how to use it. Whether writing or typing, you should photocopy the blank application form first, then complete the details in rough on the copy, before finally filling out the actual form. This will enable you not only to refine the information you present, but also to check the format: how much will fit on one line and so on. Read the whole form before you start drafting details.

As with a CV, be fussy. Take care over every detail, including simple things like the precise name of the company, of the vacancy, and of the institution at which you are studying.

Additional statement

The SAF gives the opportunity to expand on the basic facts in a number of places. Other application forms ask mainly for facts and then leave a space for "Further information in support of application" or something similar. For someone who has already started a career this tends to concentrate on previous experience. For a student it may be more about potential and aspirations.

Remember that while it is important not to appear diffident or lacking in self-confidence, there is also a danger in coming across as arrogant and pushy, or simply too good to be true. Write down what you think your potential employer should know about you, without exaggeration or padding. Whenever you can, give specific examples of achievements. Take care with use of English. The additional statement is a major test of communication skills (and will be seen that way by employers too).

Checklist for a job application

- have you made use of the careers advice facilities at your institution?
- have you researched the organisation, and thought about the sort of person they are looking for?
- have you checked whether you should apply with a CV or an application form?
- are your education details correct?
- have you covered all relevant work experience?
- have you shown yourself in the best light for this particular job?
- have you been consistent, and fussy about details?
- is your application tidy and smart?
- is it well written?
- has someone else checked it?

Further reading

There are hundreds of books on applying for jobs. Pick one (from your library) that you like the look of, or preferably follow the advice from your careers service. They will have guides and pamphlets (free or reasonably priced) specially for students.

17. Interviews

There are two parts to this chapter. The first continues the theme of job applications by concentrating on the job interview. The second is firmly back in the world of college, considering interviews that are part of the assessment of project work.

17.1 Job interviews

Being invited for a job interview means that the first stage of your job application has been a success. You should feel pleased, but it is too early to celebrate; there's a major challenge ahead.

What is the nature of the challenge of an interview? It is not to outwit the interviewers; their aim is to find out what sort of person you are, and there is no point in trying to convince them that you are someone you are not. The main challenge is to ensure that by the end of the interview there is nothing good about you (and relevant to the job) that the interviewers are not aware of. This can sometimes be difficult because of nerves or shyness or loss of concentration.

As with spoken presentations, nerves are not necessarily a bad thing as long as they don't ruin your performance. Some advisers say "the interviewers will be as nervous as you are". This may be correct on occasions, but one thing is always true: the interviewers will all have been interviewed themselves for jobs in the past, will probably be interviewed in the future, and will know what it feels like.

The best way to counteract the damaging effects of nerves, and to improve your performance generally, is to **prepare well**.

Preparation

You will have carried out a great deal of the preparation at an earlier stage of your application (as described in the last chapter). You should possess information about the vacancy and about the organisation. Part of your preparation for the interview will involve looking back at the material you collected when you made your application, and looking again at the application itself.

When preparing for the interview, in the same way as when working on the application, you should try to make full use of the careers guidance facilities at your institution. Your careers advisers will give you tips on interview technique and suggest further reading. They may have videos on being interviewed which you should find helpful. They may even put you through a mock interview.

An alternative is to get a friend to give you a mock interview. You must choose this friend carefully; you must both be able to take the exercise seriously for it to be of value.

As well as this you can practise answering questions in your imagination. You can do this anywhere: walking down the street, sitting on the bus, lying in the bath.

Asking yourself testing questions (or being asked them by a friend) can help to make you prepared for the unexpected. You might even imagine being asked some of the questions by an off-hand or aggressive interviewer, though this tactic is not particularly common. Most interviewers want to find out what you are really like, not make you more nervous. Also your interviewer might be your future boss and will not want to make an enemy of you at the start.

Here are some examples of questions. This is not a comprehensive list, it is just a starting point for you to make your own.

Interview questions

Why have you applied for this job?
What particularly interests you about it?
What makes you suitable for the job?
What are you seeking in any job?
What plans do you have for your career?
What do you see yourself doing in five (or ten) years time?
What talents/abilities will you bring to this organisation?
What are your strengths and weaknesses?
How would you describe your own personality?
What is special about you?
Give an example of your leadership abilities.
Who else are you applying to? Which job would you accept if you were offered them all?
What challenges will this organisation face in the next two years?

Many of the questions you will be asked will be based on your application form. You should reread it thinking of likely questions.

Asking yourself questions will give you practice in composing answers. You should not try to memorise an answer to any question, because it will sound unnatural when you repeat it. But phrases you have composed during one of your imaginary answers will come back to you during the interview and can be worked into your answers.

When you prepare, you should bear in mind the qualities that you think the interviewers will be looking for. Remember that these will include professional qualities like education, experience and aptitude, and personal qualities like enthusiasm and cheerfulness. The interviewers will want to find someone who will do the job well, and someone who they will get on with.

The most effective form of interview practice is attending real interviews. This is certain to improve your technique, even if you might wish that you could get the perfect job first time. Try to be positive about rejection; you will always learn something.

The interview

The interview will be with one other person or with a panel. There may be two stages: a preliminary interview, followed (for those successful at the first stage) by a second interview.

In either case the programme may include "informal" sessions with all the interviewees together, at which you will be given information on the organisation or shown round. These are difficult times: you are obviously being assessed to some extent and being compared with the other applicants, but you should try to pace yourself, and not become worn out before the interview itself. The programme will probably have been organised in this way because it is more convenient for your hosts to organise certain activities for everyone together. There is no point in trying desperately to impress to the people who show you round the building, when they may not be greatly involved in the selection in any case.

You may also have to take a series of tests, sometimes lasting more than one day. In this case it is even more important to pace yourself and to demonstrate your abilities without showing off or pretending to be something you are not.

Let's concentrate on the interview itself. You should arrive early, and should aim to eliminate any possibility of arriving flustered or disorganised. Report for the interview between five and ten minutes before time. If you get to the building much earlier, have a walk round the block, or sit in the park to compose your thoughts.

The first impressions that you make on the interviewers are important. You must be dressed smartly; you should have given this aspect particular attention, as students don't often need to be smart. The best assumption is that you cannot be too smart for an interview. As you walk into the room don't try to look artificially "dynamic", or let yourself look worried. Be cheerful and polite. If you have brought papers or samples of work, have everything well organised and ready to refer to.

Sit in an alert but comfortable position. Speak naturally: don't try to put on an artificial accent, but at the same time don't use colloquialisms or speak carelessly.

You will be asked at the end if you would like to ask any questions yourself. It is entirely appropriate for you to take up this opportunity, and you should make sure that you have found out everything important that you want to know. It is often wise not to ask about salary. (Your conditions of employment will be specified if you are offered the job.) You should make sure you fully understand the arrangements for training. Don't ask questions for the sake of it; if all your important queries have been answered during the day, say so.

Some students are a little daunted by the contrast between student life and the world of work, and fear that their interviewer may be rather scornful towards students. If this were true, the company would not be interviewing students or employing graduates. Youth, and the enthusiasm and adaptability that go with it, is an asset. You certainly don't have to pretend to be old for your years or "boring", although an employer would be put off by immaturity.

Interviews for vacation experience or sandwich placements

These are likely to be more relaxed than interviews for full-time jobs, and are unlikely to be with more than one person. The qualities that the interviewer will be looking for are commitment, initiative and flexibility. The obvious aspects of interview technique, for example smartness and

Checklist for preparing for a job interview
- have you looked back over all the material you acquired when you first made your application (especially the application itself)?
- have you thought again about the qualities the organisation will be looking for, and the qualities you have to offer?
- have you made a list of likely (and unlikely) questions, and answered them yourself?
- have you used friends and your careers service to help you prepare?
- do you know when and where the interview will be held, how to get there, and how to ensure you will arrive on time?
- are your smart clothes ready?

punctuality, are just as important as for full-time job interviews, as they show that you are serious about work.

17.2 Project interviews

Student projects are often partly assessed by interview or "viva" (short for *viva voce*, Latin for "with the living voice"). There may be important marks at stake, but in one sense these interviews are easier to cope with than job interviews. In a project interview the nature of the challenge is more easily defined; you are being assessed on fewer fronts; first impressions and selling yourself as a person are less significant.

The main purposes of project interviews are to determine:

that the work is yours
that you understand it
that you can explain it.

As with other "speaking tests" most people get nervous, but it should be possible to control nerves more in this type of interview than in a job interview or spoken presentation. The likely outcome of the interview is really determined long before the interview takes place. Either you have a good understanding of what you did in the project or you don't, and no amount of quick thinking or good performance in the interview is going to change that. So there is really nothing to be nervous about (unless you *don't* understand what you did).

Preparation

If a member of staff has given you personal supervision during your project, then that member of staff can help you prepare for the interview. To make full use of the opportunities you will need to complete your report well before the interview. Then your supervisor should be able to comment on the strengths and weaknesses of your work, and perhaps give you an idea of the types of questions you are likely to be asked in the interview.

You can prepare yourself by reading the report again and writing down likely questions. You must have your own copy of the report. While you do this you should make sure that you understand, and could explain, every detail. (Don't just revise your favourite bits.) There is no need to prepare precise answers to specific questions since, as in job interviews, a pre-prepared answer will never sound natural. You should simply think about the sorts of comments you would make.

You are likely to be asked to give a summary of what you did and what you achieved. This request is very predictable, and yet it seems to take many students by surprise. Again it is not a good idea to learn a statement by heart, because the question could take many forms, for example: "How would you summarise your project to a non-engineer?" or "What's the most important thing you found out?", or there could be an off-puttingly informal request like "Well, tell us what you've been up to".

The interview

A common format is to be interviewed by a small panel of staff, including some who have read your report and some who have not. Your supervisor will probably be there, as much to help you as to assess you. It would be worth asking your supervisor beforehand about who will be interviewing you.

Your interview will probably be fairly relaxed, but don't be put off by any apparent lack of friendliness. A few lecturers like to be stern or abrupt in these situations – they can't help it. Be cheerful yourself, and show your enthusiasm. Don't assume that a question asked in a seemingly aggressive way is a hard question, or one asked in a friendly way is an easy question. The interview is fundamentally about knowledge and understanding.

If you have a good grasp of the material, there is no reason why you should not look forward to your interview. You will be talking about something that interests you, and other people will be taking an interest in what you say. Your lecturers will have to listen to you for a change!

Further reading

See Chapter 16 (for job interviews).

18. Professional communication

The main aim of this book has been to help you write and speak clearly while you are a student. Most of what you learn will be useful to you when you become a professional engineer. However you will then need to communicate in a number of additional ways, and some of the familiar ways may need a change of emphasis.

This chapter is intended as a brief introduction to these new areas. When you need to learn more you should refer to the **Further reading** suggestions.

Spoken

Let's deal with communication in speech first. As a student, you must communicate clearly, and work well in a group, in certain fixed circumstances. But otherwise you have a great deal of freedom to work and to relate to others in the way that you choose. Also since most engineering courses still place emphasis on written examinations, some of your most important challenges take place during three hours of complete silence!

In the engineering profession you will find a more constant requirement to communicate with others. You will be working directly with other people most of the time. Much of the communication will be informal, yet must allow the communication of technical information with precise clarity.

Meetings are a more formalised type of spoken communication. Young engineers will be contributors to meetings, and as they become more senior they are likely to take on the important role of chairing.

Even when you have found a job that you are happy with you will not have seen the last of interviews. Many organisations have internal interviews for staff appraisal, salary negotiation and promotion. As you become more senior, you will find yourself on the other side of the interview table, for internal interviews or for employing new staff. Interviewing effectively is a demanding task.

Your practice in making spoken presentations as a student will benefit you when you are an engineer. You may be asked to make a presentation on a proposal to a client, or a public lecture or presentation to a conference. A specialised form of spoken presentation which may become your responsibility is the presentation of technical evidence to a court of law or a public enquiry.

Some of these forms of spoken presentation include the important component of communicating technical information to a non-technical audience. Modern society needs more engineers who are good at this.

Many business skills are forms of spoken communication. The art of negotiation is one of the most important. (See **Further reading.**)

Written

Not all written communication between engineers is formal. Notes and faxes can be informal, and yet should always be clear and precise. The slightly more formal communication within an organisation is the memo (short for memorandum; plural, memoranda). This may be on company stationery, with a standard format at the top:

to:
from:
date:
subject:

Engineers may need to spend time writing records. Meetings are recorded as detailed minutes (who said what, what was decided etc.), or in the form of summary reports. Records must be kept of progress on site, or safety procedures. Engineers involved with day-to-day running of a plant or a project keep diaries.

Many of the types of report written by engineers (and by students) have already been mentioned in Chapter 10 (Reports). There are however some fundamental differences between the writing of reports by engineers and by students.

1. When a student writes a report, the reader (the lecturer) usually knows at least as much about the subject as the writer. When engineers write reports they usually know more about the subject than the readers (that is partly why they write the report).

2. When engineers write reports they must take particular care to define their readership.

3. The readers of an engineer's report may not read the whole report. That is why it is important to write a good summary, and why it may be a good idea to place conclusions and recommendations near the start rather than the end.

4. There may be standard formats for some types of report and standard points of style required by particular organisations.

What next?

Carry on learning for yourself, and aim to continue improving as a communicator. Don't underestimate the importance of this, or you may not achieve your full potential. Professional engineers will tell you, "That's what most of my job consists of: communication".

Further reading

Scott, Bill. *Communication for professional engineers.* Thomas Telford, 1984. I recommend the sections on meetings, interviewing and negotiation. The section on spoken presentations seems better suited to confident speech-makers than technically-minded engineers.

Turk, Christopher and John Kirkman. *Effective writing: Improving scientific, technical and business communication,* 2nd edn. E.& F.N. Spon, 1989. This comprehensive book is especially good on writing-style for technical instructions and descriptions.

van Emden, Joan. *Handbook of writing for engineers.* Macmillan, 1990. This contains good advice on style and format for professional engineers.

Sides, Charles H. *How to write and present technical information,* 2nd edn. Cambridge University Press, 1992. This is a snappy American book on professional communication in computer-related fields.

Complete list of
further reading suggestions

Λ dictionary

A thesaurus

A pocket-sized book on English usage

Bryson, Bill *Dictionary of Troublesome Words*. Penguin, 1987.

Dummett, Michael *Grammar and Style*. Duckworth, 1993.

Eisenberg, Anne *Effective technical communication*, 2nd edn. McGraw-Hill, 1992.

Gowers, Sir Ernest (revised Sidney Greenbaum and Janet Whitcut) *The Complete Plain Words*. Penguin, 1987.

Haslam, Jeremy M. *Writing engineering specifications*. E. & F.N. Spon, 1988.

Huff, Darrell *How to lie with statistics*. Penguin, 1991.

Pauley, Steven E. and Daniel G. Riordan *Technical report writing today*, 4th edn. Houghton Mifflin, 1990.

Pentz, Mike and Milo Shott (ed. Francis Aprahamian) *Handling experimental data*. Open University Press, 1988.

Roze, Maris *Technical communication – the practical craft*, 2nd edn. Merrill-Macmillan, 1994.

Scott, Bill. *Communication for professional engineers*. Thomas Telford, 1984.

Sides, Charles H. *How to write and present technical information*, 2nd edn. Cambridge University Press, 1992.

Stanton, Nicki *Communication*. Macmillan, 1990.

Thirlway, Martyn *Writing software manuals – a practical guide*. The B.C.S. Practitioner Series, Prentice Hall, 1994.

Turk, Christopher and John Kirkman. *Effective communication: Improving scientific, technical and business communication*, 2nd edn. E. & F.N. Spon, 1989.

van Emden, Joan. *Handbook of writing for engineers*. Macmillan, 1990.

Answers to tests

2a

All the answers appear in Chapter 2. Here's where to find them.

1. The correct spelling of all these words is given on page 10.
2. An observable fact; phenomena; Greek. Other tricky plurals are on page 10.
3. Adverbs (and other parts of speech) are considered on page 9.
4. its/it's is discussed on page 11, the rest on page 14.
5. See page 13.
6. See page 10.

2d

Other information to support application

My **principal** reason for applying for this post is that I feel that C G Freeland is a company which can be proud of **its** record in engineering. Your new activities in **environmental** control (to which I **believe** I can contribute) perfectly **complement** your existing specialisms. I am interested in post E15 (Graduate Engineer), but would like to be considered for any **alternative** vacancies.

My main **leisure** activity is **swimming**. I have won county medals on no **fewer** than six **occasions**, and I represented England at the last Commonwealth Games.

3a

This is a piece of writing with no punctuation. What you must do is insert it. You may use commas, full stops, capital letters, paragraphs, or any other forms of punctuation that you think might be appropriate.

The main unit of written English is the sentence. Sentences can be long or they can be short. A sentence really expresses one thought. If the thought can be expressed in a brief statement, it is quite appropriate for the sentence to be short. However some thoughts are more complex, and are linked together by words like "since", "and", "because" or "but". In technical English, where clarity is of prime importance, there is more danger in long sentences than short ones. Of course every sentence must contain a verb. If it doesn't, it is (in a manner of speaking) too short.

Paragraphs are also very useful for bringing clarity to written English. The break between paragraphs provides a definite pause in the text.

Punctuation really matters because it helps to make writing clear. People who do not write clearly, perhaps because their punctuation is poor, make a bad impression. They may be excellent at their profession in other respects, but if they fail to get good jobs, or fail to gain their clients' trust, they are likely to be disappointed in their careers.

3c

1. The number of members within a team depends on two factors: the size and complexity of the project.
2. A quality management system should be based on existing systems, amended and supplemented where necessary to conform with BS 5750.
3. Control should be exercised throughout the whole process from start to finish; products within a subcontractor's work may have to be included.
4. What are the main problems with the current system?

3d

1. The electronics industry has been healthy compared with other industries; this can be clearly seen in the attached graphs.
2. This, coupled with high interest rates, has caused many small engineering firms to fold.
3. Engineering will continue to be misunderstood, and we graduate engineers are the ones who will suffer most. (second comma removed)
4. The wall is relatively thin, but it is strengthened at regular intervals by buttress supports. (or remove comma)
5. Although these machines rarely need maintenance, do not have regular breaks like their human counterparts, and do not arrive late, they were not developed to replace humans.
6. Another interesting idea is one that is currently used in Houston, Texas. (first comma removed)
7. There is no requirement for the engineer to be present. Isn't this unsatisfactory?

10a

Here is a possible answer.

1. Introduction
   (client, address of property, terms of reference)

2. Accommodation
   2.1   Summary (list of rooms that can be used)
   2.2   Room details (size, access, person-capacity, features)
   2.3   Other features (hall, stairs, balcony, garden . . .)

3. Facilities
   3.1   Kitchen (for food preparation, washing-up . . .)
   3.2   Sound equipment

4. Neighbours
   4.1   Within property (possibility for shared event, noise)
   4.2   Neighbouring property (noise limits)

5. Transport
    5.1 Parking
    5.2 Public transport
        5.2.1 Closeness to route(s)
        5.2.2 Timetable (frequency, last departure)
    5.3 Taxi fares

6. Recommendations

10b

There is no single right or wrong answer. I suggest this:

1. Introduction
    date of meeting, purpose
    number of students attending

2. Course content
    2.1 General **a. i. l.**
    2.2 1st year **c. r.**
    2.3 2nd year **b. h.**
    2.4 Final year **t.**

3. Staff **j. e. q.**

4. Academic facilities
    4.1 Teaching rooms **d.**
    4.2 Laboratories **m. n.**
    4.3 Computing **f. g. k.**
    4.4 Library **o. p.**

5. Catering **s. u.**

6. Comment
    the meeting was successful
    hope the points are useful for course planning

10c

This is my attempt.

"Engineer" has the same Latin root as "ingenious". Engineers cannot put their ingenuity into practice unless they can communicate well. Poor communication can damage reputations, even cause disasters. Yet employers find graduate engineers generally poor at communicating. It is felt that engineering courses should play a part in remedying this.

Good engineering and good communication have a lot in common. They both require judgement, not simply application of fixed rules. Engineers must be clear when they communicate: clear about *what* they are communicating and *how* they are communicating it. Tables and diagrams can enhance their words.   (96 words)

Index

For words in *italics*, the text gives advice on use of the actual word.